Minimally Invasive Neurosurgery I

Edited by
B. L. Bauer and D. Hellwig

Acta Neurochirurgica
Supplementum 54

Springer-Verlag Wien GmbH

Professor Bernhard Ludwig Bauer, M.D.
Dieter Hellwig, M.D
Department of Neurosurgery, Philipps-University Marburg, Federal Republic of Germany

With 113 partly coloured Figures

Typesetting: Thomson Press, New Delhi, India

ISSN 0065-1419
ISBN 978-3-7091-7381-7   ISBN 978-3-7091-6687-1 (eBook)
DOI 10.1007/978-3-7091-6687-1

Preface

The first International Workshop "Contemporary Update on Endoscopic Neurosurgery and Stereotaxy" held in Marburg 1990 enjoyed a widespread and an encouraging response. At that time four years had elapsed since *Hugh B. Griffith* published his survey of the history and the contemporary standard of endoneurosurgery in *Advances and Technical Standards in Neurosurgery*.

Efforts to introduce the endoscope into the practice of neurosurgery are not new. As early as 1922 *Walter Dandy* reported on his first ventriculoscopies carried out on patients with hydrocephalus. In many other medical and surgical fields the advantages of endoscopic procedures have been recognized and accepted. In the field of neurosurgery new points of view have been developed over the past few years and techniques of minimally invasive procedures have been refined. Technical advances therefore demand a new anatomic-endoscopic understanding and a change in our strategy in operative approach. Highly sophisticated microinstruments, laser equipment and ultra-sound guidance have converted the endoscope from a purely "visualizing instrument" into a "surgical instrument". Endoneurosurgery has been pushed towards the forefront of neurosurgical interest.

An element of overestimation often accompanies the invention or discovery of new medical techniques or surgical approaches. Enthusiasm for these "certain key to success" techniques must therefore be tempered by intelligence and caution. Unresolved problems require the provision of well-organized workshops and meetings. Neurosurgery founded on the principles of the so-called keyhole surgery is not an end in itself. The development and advances from macrosurgery to microsurgery is largely complete. The next step to minimal invasive neurosurgery requires besides these new techniques new anatomical and neurophysiological understanding about intracranial space-occupying lesions. Such consideration is to be commended, not least because medical history provides many examples of failed surgical approaches and techniques.

The actual starting point of our own involvement in this field was, some years ago, the simple desire to increase the safety of stereotactic bioptic procedures by directly visualizing potential sources of bleeding. The method quickly proved its further worth in the treatment of cystic space-occupying lesions and in the evacuation of chronic extracerebral and acute intracerebral hematomas. It is absolutely clear that these new endoscopic procedures must be evaluated according to the standards of classical neurosurgical approaches. Technical assessment of these innovations is a necessary ongoing project. What is presently lacking is a clearcut list of indications for endoneurosurgical procedures. Contrasting views of various authors make if difficult to come to a concensus. Nevertheless it seems possible to be able to give a modest list of basic recommendations for the use of endoscopic procedures in neurosurgery. For these procedures we chose the term *minimally invasive neurosurgery (MIN)* with respect to the term *minimally invasive surgery* first defined by *J. M. Fitzpatrick* and *J. E. A. Wickham*, 1990.

The papers presented in the first workshop were an attempt to combine and present such an assessment to neurosurgeons active in this field. This volume is neither the first nor the last stepping stone on the path towards a fully integrated endoneurosurgery. We hope that the efforts we made in this special field may encourage neurosurgeons to demonstrate and to discuss their own experience. The sooner these papers presented in this volume are discussed, the greater their success will be.

Marburg, December 1991

The Editors
Bernhard L. Bauer, Dieter Hellwig

Contents

Acta Neurochirurgica, Suppl. 54, 1–10 (1992)

Topographic Anatomy of Preformed Intracranial Spaces

J. Lang

Department of Anatomy, University of Würzburg, Federal Republic of Würzburg

Summary

This article gives descriptions and measurements of the cerebral ventricles, especially our measurements of the interventricular foramen and the third ventricle. Included are measurements of previous and recent research. The results of endoscopic reviews of the lateral, third and fourth ventricles are also discussed.

The subarachnoid spaces are described and illustrated by our corrosion casts.

During endoscopic inspection of the subarachnoid spaces, the transcisternal veins are extremely vulnerable. Therefore, these veins in the anterior, middle and posterior cranial fossae are described.

Keywords: Cerebral ventricles; location and measurements; Subarachnoid spaces; Transcisternal veins of the anterior, middle and posterior cranial fossae.

Introduction

The cerebral ventricles and the subarachnoid cisterns, the cerebral arteries and veins as well as the dural sinuses may be regarded as preformed intracranial spaces for endoscopic and, partly, also for endoneurosurgical operations. Therefore, I would like to represent briefly the ventricular system, the subarachnoid spaces, some cerebral vessels and also take a look into the spinal canal. Especially it should be pointed to dangerous zones by the endoscopic access.

1. Ventriculi Cerebri (Cerebral Ventricles)

a) Ventriculus Lateralis

In the current Nomina Anatomica (1989) a pars centralis is distinguished from the cornua frontale, occipitale and temporale. According to most of the researchers, the left lateral ventricle, especially its occipital horn, is larger than the right. Particularly Torkildsen (1933/34) gave distances of the different ventricular parts to the brain surface on the basis of numerous measurements as well as ventricular measurements after pneumo-encephalography. Accordingly, the frontal horn of the lateral ventricle has a distance of 4 cm on average to the brain surface, the distances of the pars centralis are slightly shorter, those of the ventricular triangle to the border of the longitudinal fissure of the cerebrum are clearly longer. Figure 1 shows his measurements concerning the fourth ventricle in $\bar{x}$ cm and ours in $\bar{x}$ mm.

Cornu frontale (frontal horn): The roof segment of the anterior cornu is formed by the corpus callosum fibers coated with ependyma. It has a breadth of approx. 1.5 cm. The angle being formed by both roof segments is 126°. The head of the dentate nucleus indents the anterior cornu from laterally and below. The distance between the anterior border of the cornu frontale and the midpoint of the interventricular foramen was determined to be 3.2 (2.9–4.2) cm. The medial border of the anterior horn is formed by the septum pellucidum, possibly also by the lateral wall of a cavum septi, as well as by the rostrum corporis callosi (anterior end of the corpus callosum). The anterior wall itself consists of fibers of the genu corporis callosi. Trauma to these structures should be avoided; especially also to the superficial subependymal veins of the caput nuclei caudati as well as anterior veins of the septum pellucidum. These vein segments covered by thin ependyma are particularly vulnerable.

In our material the *interventricular foramen* has a long diameter of 5.1 (2–8) mm and a short one of 2.9 (1–6) mm. Its anterior and upper circumferences are formed by the fornix. This fiber strand being part of the fiber complex of the limbic system contains approx.

2.7 million fibers, which take a cortico-afferent and cortico-efferent course. Lesions here are said to result in impairment or loss of recent memory. The anterior tubercle of the thalamus is the posterior and lateral border of the interventricular foramen. In the area between the anterior tubercle of the thalamus and the interventricular foramen is found, as a rule, the confluence of anterior subependymal veins. This is the boundary between the frontal cornu and the central part of the lateral ventricle. Medially the choroid plexus of the lateral ventricle is in direct connection with the choroid plexus of the third ventricle behind the interventricular foramen. This is the area, where colloid cysts and possibly tumours developing from the subfornical organ can impede or make the drainage of CSF to the third ventricle impossible.

According to different authors, the body of the lateral ventricle (*pars centralis*) measured from the midpoint of the interventricular foramen to its transition into the ventricular trigone, is approx. 4 (3.2–4.9) cm in adults, its greatest height is 6.5 mm on average. Anteriorly this ventricular segment is approx. 10 mm high. Its cross-section is roughly triangular in shape. The roof is formed by the under surface of the corpus callosum, the concave floor segment descends from lateral to medial and is laterally formed by the corpus nuclei caudati, then medially by the stria terminalis and the thalamo-striate vein and then apparently by a lateral part of the surface of the thalamus: lamina affixa. The medial wall is formed by the fimbria fornicis and the fornix and above it by the septum pellucidum. Endoscopically the choroid vein and the glomus choroideum at the choroid plexus are clearly visible. The choroid plexus of the pars centralis (body) is attached to the fimbria fornicis and the taenia choroidea of the thalamic surface. Tearing these plexus attachments produces an entry into the cistern of the transverse fissure of the cerebrum. Twigs of the rami choroidei posteriores mediales, also supplying the surface of the thalamus, run through the fissure and also to the plexus. In addition, a so-called posterior drainage of the thalamo-striate vein can be present dorsal to the interventricular foramen (in our specimens in 20%). Attention has to be paid to this vein. In the roof segment of the body of the lateral ventricle the vena atrii medialis intersperses the fornix, as a rule, and reaches the cistern of the transverse fissure underneath the splenium corporis callosi. The vein of the lateral atrium arises from the lateral wall of the atrium and courses from the occipital cornu at the same level medial to the most dorsal segment of

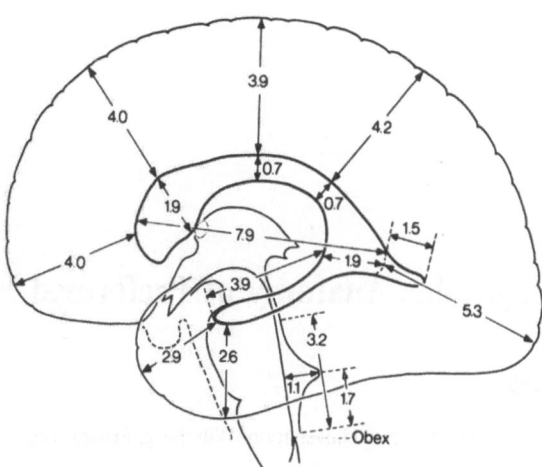

Fig. 1. Ventriculi cerebri, length and height measurements as well as distances to different areas of the surface of the brain (according to Torkildsen, 1933/34). The measurements are given in centimetres and those of the fourth ventricle in millimetres (own material – Deymann-Bühler, 1984)

the stria terminalis, mostly below the stria and the choroid plexus in order to reach the cistern of the transverse fissure and the internal cerebral vein (of Galen) or (less commonly) the basal vein (of Rosenthal).

Atrium: The term 'atrium' = *ventricular triangle* = forecourt = the trigone of the lateral ventricle is forgotten in the current Nomina Anatomica (1989). The atrium (trigone) is the transition zone between the pars centralis, cornu temporale and cornu occipitale of the lateral ventricle. Here, the choroid plexus of the lateral ventricle seems to be swollen (especially in old people) to form the glomus choroideum, and this is partly calcified. The length of the atrium was ascertained to be approx. 23 mm (Taveras and Wood 1964) (see Fig. 1). Measurements of Hussein and Woischneck (1990) show variations and mean values (Tables 1 and 2). The roof is formed by the corpus callosum, its floor by the collateral eminence and trigone, and laterally by the stria terminalis and the tail of the caudate nucleus. The lateral wall is the tapetum and the optic radiation. The cornu occipitale is 1.45 cm long on average, occasionally it is absent, sometimes it has a length of 3.6 cm in earlier descriptions. The initial part of the posterior horn is mostly situated at the level of the splenium corporis callosi. Its cross-section has a convex bend laterally and so concave medially. The roof is formed by the splenium corporis callosi, the lateral wall chiefly by the visual pathway, and the medial by the corpus callosum again. Here the calcar avis projects into the posterior horn. Above from it is

Table 1. *Topographic Measurements.* (n = 100) (according to Hussein and Woischneck, 1990)

Occip. wall (posterior horn)↔ant. wall of atrium	17.2 (14–20) mm
Occip. pole↔posterior horn	41.6 (21–57) mm
Lat. thalamus↔lat. wall of ventricle	3.1 (0–6) mm

Table 2. *Structures at the Anterior Wall of the Atrium.* (n = 100) (according to Hussein and Woischneck, 1990)

Hippocampus and fornix	46%
Fornix and thalamus	32%
Fornix, hippocampus and thalamus	7%
Hippocampus	15%

a further bulge of corpus callosum fibers, which is termed the bulbus cornu occipitalis.

The choroid plexus of the lateral ventricle turns downwards and laterally into the temporal horn (*cornu temporale*) of the lateral ventricle. Its length is 4 cm on average in adults, measured from the ventricular triangle to its tip. The choroid plexus and its attachments are located at the medial side of the temporal horn. Its cross-section is roughly four-sided. The nearest brain area is the superior temporal sulcus. According to Renella (1989) the sulcus is 15–20 mm deep. At the transition zone to the posterior horn, the inferior horn is dented in by the trigonum collaterale (the collateral sulcus). The longitudinal elevation continuing to the front above the middle part of the collateral sulcus is termed eminentia collateralis. From below the hippocampus bulges into the inferior horn. The tip of the inferior horn (frequently partitioned off from the main space by adhesions) is located 8–12 mm behind the temporal pole. Anterior and medial to the tip, the tuberculum amygdalae is present as a bulge of the corpus amygdaloideum. The roof of the inferior horn and the lateral wall are constituted by the tapetum whose posterior part is the optic radiation. When the choroid plexus is removed, the opening is termed the hippocampal fissure. The inferior choroid vein and other ventricular veins project from the plexus into the inferior horn. When the choroid fissure is opened, the anterior choroidal artery, the posterior cerebral artery and some of their branches as well as the basal vein (of Rosenthal) can be reached.

b) Third Ventricle

Figures 1–4 give measurements of the inner CSF spaces (according to results of previous researchers

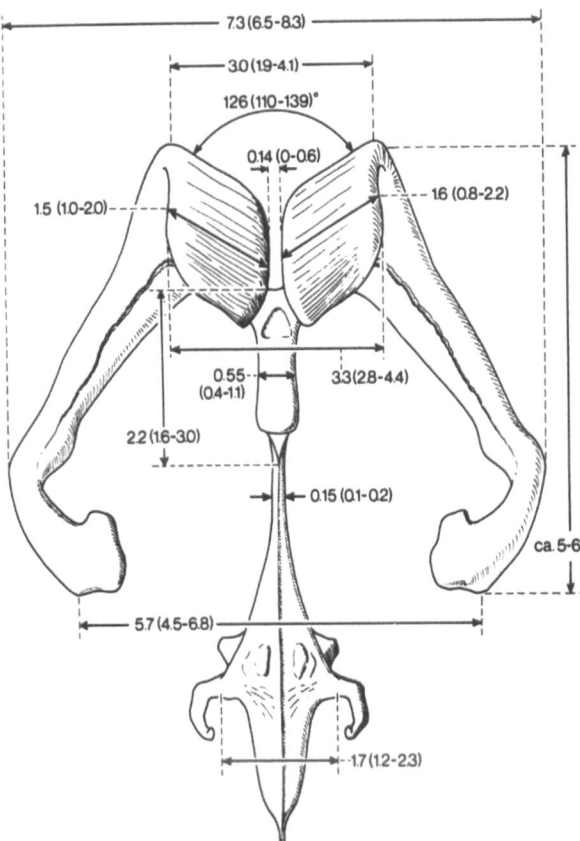

Fig. 2. Measurements of the ventricles in a corrosion cast (anterior aspect). Widths and angles are given in millimetres and degrees (according to Torkildsen, 1933/34)

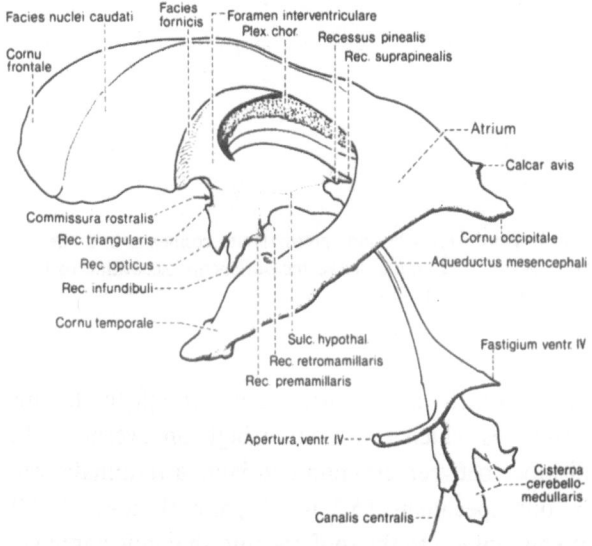

Fig. 3. Ventricles of the brain, lateral view of a corrosion cast

and our own material). Figure 5 shows the summary of some measurements in our material (Lang *et al.* 1983). The third ventricle can be endoscopically inspected by means of angled lens systems through relatively wide interventricular foramina. It should be stressed that

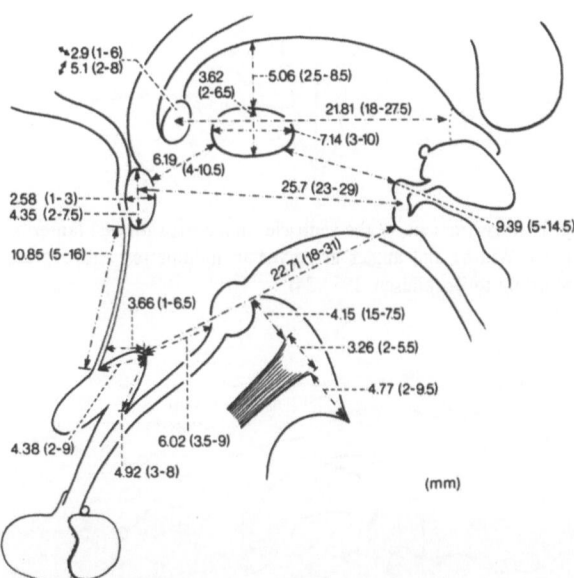

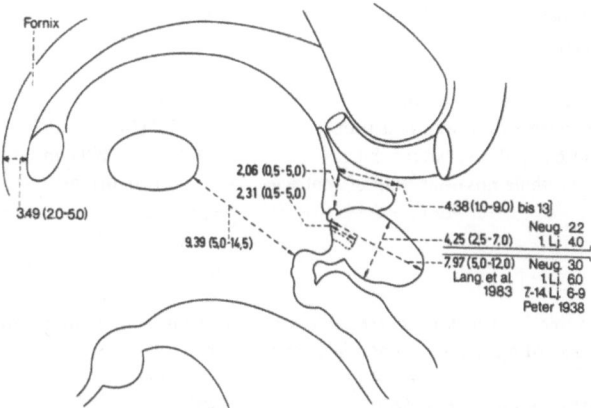

Fig. 4. Measurements of single distances of the lateral ventricles, the third ventricle and the fourth ventricle (according to different authors and Krauss, 1987)

Fig. 6. Measurements at the posterior part of the third ventricle (suprapineal recess, pineal recess), dimensions of the pineal body and distance between interthalamic connexus and commissura epithalamica (posterior) as well as measurement of the width of the fornix

Fig. 5. Third ventricle and interthalamic connexus and the exit zone of the third cranial nerve, measurements (according to Lang et al. 1983; Krauss 1987)

the approach can be carried out through the lamina terminalis which is 10.85 mm high on average. The distance between cranium and lamina terminalis was in our specimens 55.2 (49–62) mm (Krauss, 1987). When looking to the roof, the interthalamic connexus has to be by-passed. This substantial bridge was found in our cadavers in 75%. Its average length is 7.14 (3–10) mm (300 hemispheres, Lang et al. 1983), its width 3.6 (2–6.5) mm. Rarely both diameters of the connexus have a length of 10–12 mm. Small vessels also pass through it. In addition we detected transverse fibers as well as nuclear areas of the paraventricular

nucleus on the connexus. Figures 5 and 6 show its variable position in relation to the roof and the floor of the third ventricle. In addition, the distance between the anterior commissure (rostral) and the adhesion is noted. The depth and breadth of the optic recess, infundibular recess, pineal and suprapineal recesses as well as the size of the corpus pineale are shown in Figs. 5 and 6. The greatest part of the lateral wall of the third ventricle belongs to the hypothalamic regions. Close under the ependyma important hypothalamic nuclei are located as well as pathways which fulfill particular regulating tasks. Also parts of the limbic system (pars tecta fornicis a.o.) are lying under the side walls of the third ventricle. Sometimes central branches of cerebral arteries supplying the hypothalamic and thalamic regions, shimmer through the side walls. These vessels have to be preserved. As a rule, from the roof of the third ventricle two longitudinal strips of the choroid plexus project into the third ventricle. In most cases they extend into the suprapineal recess in the roof itself or in more lateral segments of the roof. The ependymal roof is attached to both striae medullares thalami which can be followed from the interventricular foramen to the commissura habenularum. It is known that twigs of the ramus choroideus posterior medialis supply the diencephalon also from above. Occasionally branches of this artery are visible through the roof of the third ventricle. The internal cerebral veins with their lateral tributaries and from the upper surface of the thalamus run above the roof in a paramedian position. Above the suprapineal recess, the position of the confluence of the internal cerebral veins is 35.3 $\bar{x}$ mm posterior

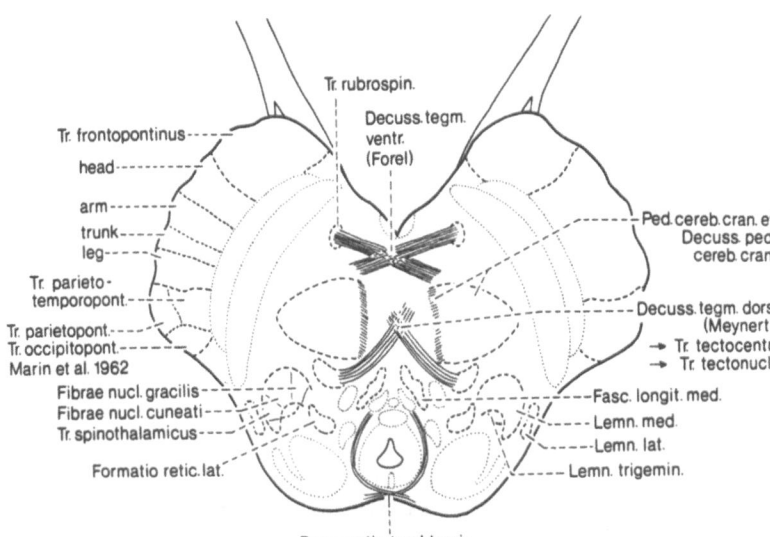

Fig. 7. Transverse-section through the caudal part of the mesencephalon at the level of the tracts and exit of the trochlear nerves

of the anterior rim of the interventricular foramen, but up to 6 mm in front or 9 mm dorsocranially behind the splenium corporis callosi (Lang *et al.* 1983).

c) Aqueductus Mesencephali (Aqueduct of Sylvius)

When looking inside the third ventricle posteriorly, there are visible from above downwards: the suprapineal recess, the commissura habenularum, the pineal recess, the commissura epithalamica and below it the entrance to the aqueduct of the midbrain. This narrow isthmus for the passage of CSF is 16.1 (12–21) mm long between the inferior circumference of the commissura epithalamica and the lower rim of the decussation of the trochlear nerves. At its cephalic end the aqueduct of Sylvius has a diameter of approx. 1 mm, then follows a slight widening, afterwards a further narrow zone (Turnbull and Drake 1966). The aqueduct forms a smaller angle with the floor of the third ventricle in children under 5 years of age than in adults. According to Lindgren and DiChiro (1953) more curved courses of the aqueduct occur in approx. 37%. Figure 8 provides information about nuclear and fiber areas in the vicinity of the aqueduct. Figure 9 shows the aqueduct with neighbouring structures in a median sagittal section.

d) Fourth Ventricle

In our specimens the fourth ventricle has a length of 32.32 (25–39) mm measured from the decussation of the trochlear nerves to the obex. Its height (measured vertically) to the fastigium is 11.36

(8–16) mm (see Fig. 10). In addition, Deymann-Bühler found the distance between the fastigium and the obex to be 16.56 (9–24) mm. The caudal portal of the fourth ventricle is the apertura mediana ventriculi quarti. This aperture is most commonly small and high, less commonly broad and low. Less commonly it is closed by fiber systems of the pia mater, or both posterior inferior cerebellar arteries are closely behind the aperture. The longitudinal extension of the choroid plexus of the fourth ventricle projects through the aperture 6–10 mm (to a maximum of 25 mm) in 40% of our specimens (see Figs. 163a–163f in Lang 1979). The floor of the fourth ventricle resembles a lengthways disposed rhomboid with small lateral segments projecting down- and forwards. The narrow lateral segments are termed the lateral recesses of the fourth ventricle. Figure 11 gives information about the nuclear and fiber areas of the fourth ventricle. In addition to the lateral recess, the fourth ventricle has a postero-superior recess and a postero-lateral recess as well as a bulging of the dentate nucleus, termed the eminentia nuclei dentati. (For further particulars see Lang 1985 from p. 417 onwards). The choroid plexus of the fourth ventricle can be subdivided into superior and inferior longitudinal divisions, and two transverse divisions to and through the lateral apertures. In the median plane there is mostly a rhomboid field without any plexus. Apart from the organum subfornicale, the organum vasculosum laminae terminalis, the organum subcommissurale, the organum paraventriculare and the area postrema, an organum recessus lateralis ventriculi quarti was also described in its roof area. For details see Lang (1985; pp. 420 ff).

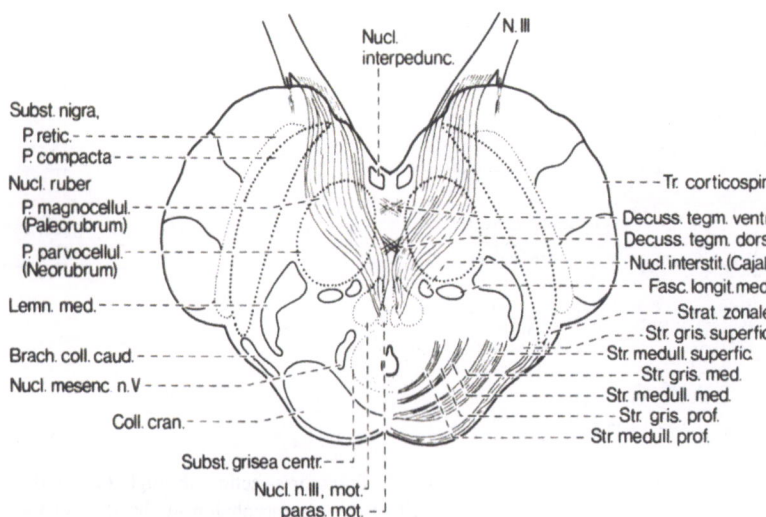

Fig. 8. Transverse-section through the mesencephalon at the level of the red nuclei

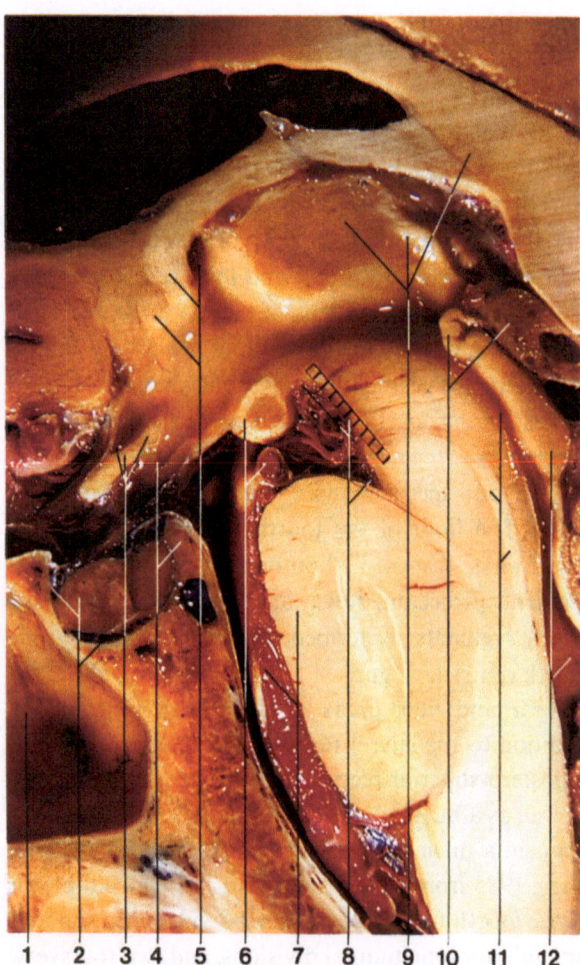

Fig. 9. Median sagittal section through the mesencephalon and its relations. *1* Sphenoidal sinus, *2* adenohypophysis and anterior and inferior intercavernous sinuses, *3* optic recess and chiasma, *4* recessus infundibuli and dorsum sellae, *5* fornix, interventricular foramen and rostral commissure, *6* mamillary body and posterior cerebral artery, *7* basilar artery and pons, *8* interpeduncular fossa, mm-strip and interpeduncular twigs, *9* stria medullaris thalami and corpus callosum, thalamus (cut), *10* commissura epithalamica and pineal body, *11* aqueductus mesencephali and fasciculus longitudinalis medialis, *12* Lamina tecti and fourth ventricle

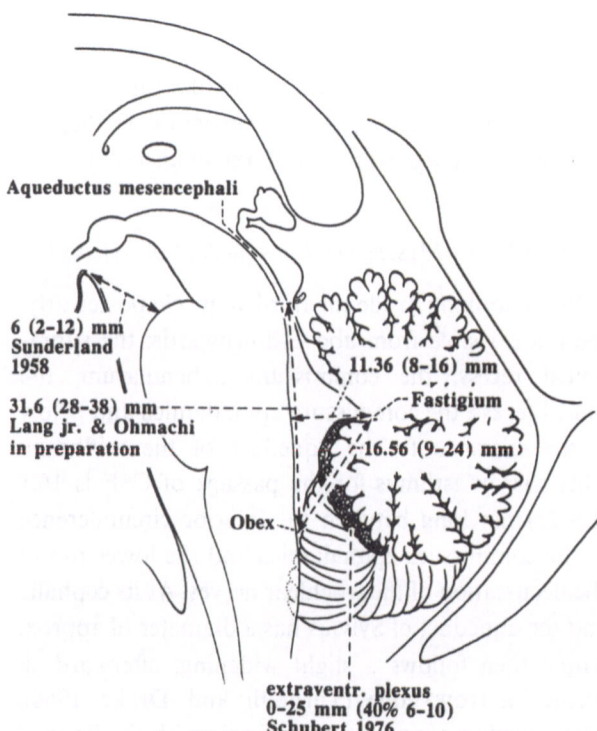

Fig. 10. Fourth ventricle and its relations

2. Subarachnoid Spaces

Described in simplified terms, the CSF circulates around the brain, its afferents and efferents as well as the intracisternal cranial nerves between the arachnoid and the pia mater. Numerous arachnoidal trabeculae pass through this subarachnoid space and produce connections to the adventitia of the vessels. Larger subarachnoid spaces dip as cisterns into greater fissures and niches of the brain surface. In adults, approx. 25 ml CSF are within the intracranial subarachnoid

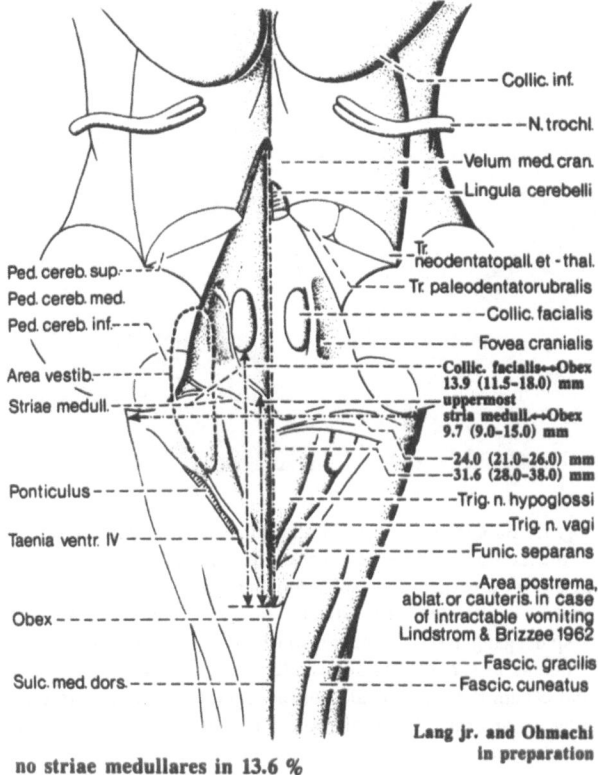

Collic. inf.

N. trochl.

Velum med. cran.

Lingula cerebelli

Tr. neodentatopall. et -thal.

Tr. paleodentatorubralis

Collic. facialis

Fovea cranialis

Collic. facialis↔Obex 13.9 (11.5–18.0) mm
uppermost stria medull.↔Obex 9.7 (9.0–15.0) mm

—24.0 (21.0–26.0) mm
—31.6 (28.0–38.0) mm

Trig. n. hypoglossi

Trig. n. vagi

Funic. separans

Area postrema, ablat.or cauteris. in case of intractable vomiting Lindstrom & Brizzee 1962

Fascic. gracilis

Fascic. cuneatus

Ped. cereb. sup.

Ped. cereb. med.

Ped. cereb. inf.

Area vestib.

Striae medull.

Ponticulus

Taenia ventr. IV

Obex

Sulc. med. dors.

Lang Jr. and Ohmachi in preparation

no striae medullares in 13.6 %

Fig. 11. Fourth ventricle, floor with important nuclear areas

spaces and cisterns, and approx. 75 ml CSF within the spinal portion (Schaltenbrand 1955). The main task of the CSF is to keep the brain suspended. All the edges and rims of the dural sheaths are upholstered by variable CSF cushions. Figure 12 gives an overview of the outer CSF spaces (and of the fourth ventricle) being most commonly found in our material in

corrosion preparations. On the left, the cisterns are removed. Especially Magendie (1822), Koelliker (1850), Key and Retzius (1875), Liliequist (1959) and Yaşargil et al. (1984) have thoroughly worked on the extensions and compartments of the cisterns.

a) Cisterna Cerebello-medullaris (Magna)

The cerebellomedullary cistern is suitable for an endoscopic approach; it has a depth of 21 (15–30) mm in the median plane between the dura of C_0–C_1 and the median aperture of the fourth ventricle. It is ventrally bounded by the medulla oblongata and the posterior wall of the fourth ventricle. Dorso-caudally its outer arachnoid lies flat on the dura at the squama of the occipital bone. Above the aperture of the fourth ventricle are the medial surfaces of the cerebellar tonsils, the lower vermis and parts of the posterior inferior cerebellar artery and its branches. Also inferior arteries of the vermis are visible from the dorsal aspect. An upward continuation is termed the cisterna valleculae. It directly turns into the differently developed inferior cerebellar cistern. The cistern becomes shallower laterally. Altogether its breadth is 3–3.5 cm, as a rule. In addition to the trabeculae arachnoideales a perforated septum between the right and left sides not uncommonly exists inside the cistern. Veins from the tonsillar and inferior vermis regions as well as of the choroid plexus of the fourth ventricle also run through the cistern. In our material it was found that in approx. 40% the cistern is traversed by

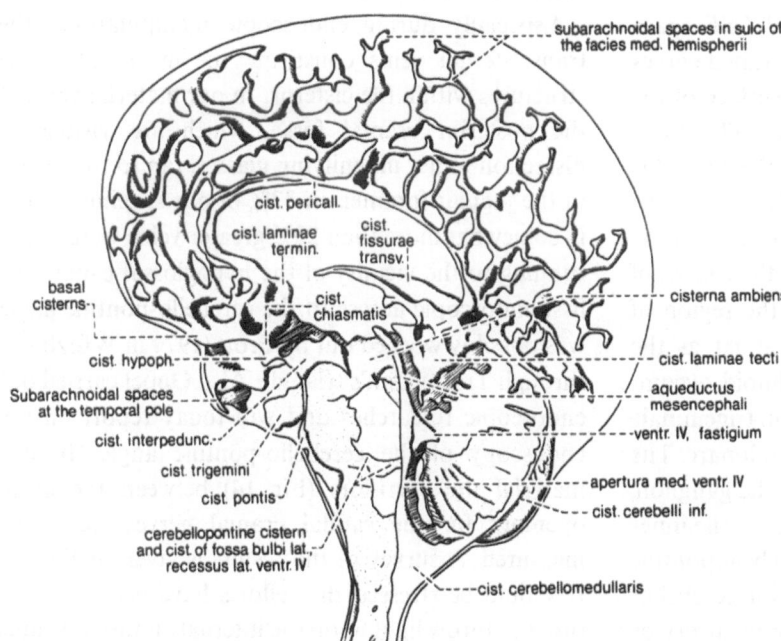

subarachnoidal spaces in sulci of the facies med. hemispherii

cist. pericall.

cist. laminae termin.

cist. fissurae transv.

basal cisterns

cist. chiasmatis

cisterna ambiens

cist. hypoph.

cist. laminae tecti

Subarachnoidal spaces at the temporal pole

aqueductus mesencephali

cist. interpedunc.

ventr. IV, fastigium

cist. trigemini

apertura med. ventr. IV

cist. pontis

cist. cerebelli inf.

cerebellopontine cistern and cist. of fossa bulbi lat. recessus lat. ventr. IV

cist. cerebellomedullaris

Fig. 12. Subarachnoid spaces of the brain and the cranio-cervical junction (after an injection preparation). On the left side the CSF spaces are removed in the median plane

a vein from the far side of the medulla oblongata which runs to the occipital sinus. Rarely branches of the posterior inferior cerebellar artery (PICA) course through the cistern to the squama ossis occipitalis and to the dura mater. These are posterior meningeal arteries. When looking into the spinal CSF space, the dorsal root fibers of the cervical nerves, the spinal accessory nerve and the radicular arteries and veins are discernible dorsal to the denticulate ligaments (Fig. 13). The root fiber bundles run to the entrances of the root pockets, actually in the cervical region to the ventro-lateral aspect of the spinal subarachnoid space. The portal for the vertebral artery and the uppermost serration of the denticulate ligament are visible. When the endoscope is moved more laterally, the ventral root fibers and their accompanying vessels are seen. The posterior subarachnoid space of the upper spinal cord is mostly incompletely divided into two subdivisions by the septum posticum. Further and more uncommon attachments of the spinal cord have been described (Lang and Emminger 1963).

Between the medulla oblongata and the pons a further arachnoid layer exists which is incompletely closed to the pia mater and the basal brain vessels, and represents the demarcation between the cerebello-medullary cistern and a lateral cerebello-medullar cistern (Lang 1979; Yaşargil et al. 1984). The arachnoid leaf stretches over the nn. IX–XI and separates this cistern from the cerebello-pontine cistern.

b) Cisterna Cerebello-pontina (Cistern of the Cerebello-pontine Angle)

A truncated and three-sidedly defined CSF space extends into the niche between the pons, cranial nerves VII and VIII on the one hand and the surface of the inferior olive on the other hand (Lang 1972, 1973; Lang and Nadjmi 1973). Lateral to it the flocculus forms the dorsal boundary. Cephalically, the cistern merges into the arachnoid of the ambient cistern. Laterally the cisternal wall continues to the cistern of the internal auditory meatus as far as the region of the fundus and along the facial nerve as far as the geniculate ganglion. Above it the arachnoid accompanies the trigeminal nerve into the cavum trigeminale (Meckel's cave) as far as the ganglion semilunare. The cistern is fixed there on the outer side of the ganglion, while a niche-shaped CSF space is found on the inner side of the ganglion. Within the cerebellopontine cistern run, as a rule, the anterior inferior cerebellar artery (AICA) and the internal labyrinthine artery or

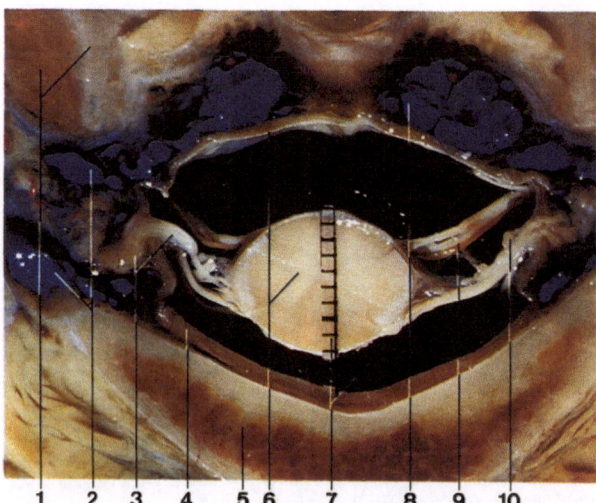

Fig. 13. Subarachnoid space of the spinal canal (transverse section at the level of the upper segment of the axis), inferior aspect in the case of an especially wide subarachnoid space. *1* Inferior articular process of the atlas with articular surface, *2* intervertebral veins, *3* spinal ganglion of C$_2$ and dorsal root of C$_2$, *4* dorsal dura mater, *5* lamina axis, *6* ventral dura mater and spinal cord, *7* mm-strip and cavum subarachnoidale dorsale, *8* plexus venosus vertebralis internus anterior, *9* ventral root of C$_2$, *10* spinal portion of the accessory nerve

arteries, the subarcuate artery and transcisternal veins. Most commonly the outflow of the petrosal vein (Dandy's vein) is lateral to the trigeminal cistern, less commonly medio-caudal or caudal from it. In 10% two veins of that kind occur, less commonly three draining to the superior petrosal sinus. The confluence site is approx. 24 (17–39) mm paramedian and has a distance of 27 (11–38) mm to the lateral border of the cerebellum (Lang and Debes 1977).

Especially during endoscopic manipulations, the transcisternal veins constitute the most vulnerable structures within the cisterns. In our material veins of the posterior cranial fossa, which are visible by dissection under magnifying glasses, ran to the region of the jugular foramen in 32% of our specimens, and in somewhat more than 23% greater veins were found running to the vicinity of the hypoglossal canal. The first endoscopic access to the cerebello-pontine angle was possibly worked out by Prott (1974 in Würzburg) through Trautmann's triangle. Also Oppel carried out endoscopic researches and will today report on the endoscopy of the cerebello-pontine angle. In our material the corridors (Fig. 14) between the dural openings for the caudal cranial nerves were also measured. A survey of the results is given in Fig. 15. It should be stressed that efforts have been made in order to throw light upon the internal auditory meatus

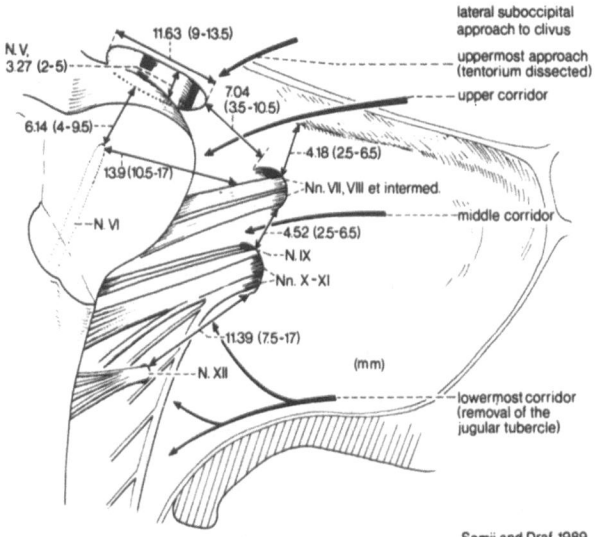

Fig. 14. Corridors between the entries of the caudal cranial nerves towards the apex partis petrosae, clivus a.o.

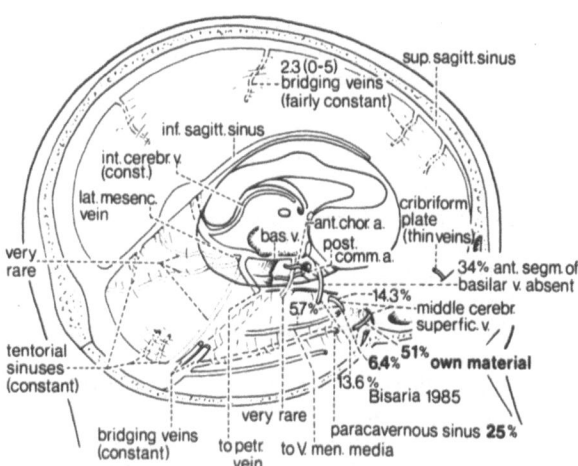

Fig. 15. Bridging veins, detected in our material, to the falx cerebri, lamina cribrosa, to the region of the cavernous sinus as well as to the upper genu of the sigmoid sinus, transverse and straight sinuses

endoscopically in different ways (medially, by the translabyrinthine or retrolabyrinthine approach).

c) Other Endoscopic Accesses and Danger Areas

On the subdural way between the dura mater and the arachnoid can also be proceeded endoscopically. The requirement for that is a knowledge of normal anatomical conditions and variants as well as pathological conditions in the patients. Currently these can be cleared up with the aid of modern diagnostic techniques (CT, MRI a.o.). It should be pointed out that in our material, although uncommonly, branches of the anterior cerebral artery to the lamina cribrosa and veins from the under-surface of the frontal lobe to this area were found. In my opinion, particular importance attaches to the vulnerable veins in the subarachnoid and subdural spaces.

Figure 15 shows the drainage of the v. media superficialis cerebri to the cavernous sinus, the sphenoparietal sinus, into a paracavernous sinus as well as less common courses, for instance as bridging vein to the upper curve of the sigmoid sinus. In 34% an anterior drainage of the basal vein of Rosenthal flows into the anterior side wall of the cavernous sinus. Scarcely a vein of that kind runs into the posterior segments of the side wall of the cavernous sinus and between the anterior choroidal and posterior communicating arteries. Drainage of the basal vein to the petrosal vein (vena mesencephalica lateralis, lateral mesencephalic vein) was found in 70% in our material (Lang *et al.* 1987). This vein is 0.81 (0.29–1.78) mm

wide. Rarely drainage of the basal vein was detected under the tentorium cerebelli to the straight sinus or to the confluence of the sinuses posteriorly. Drainage to the sigmoid sinus underneath the tentorium cerebelli was also found. We term those vein segments, whose one wall takes root on the dura mater, bridging veins. Such bridging veins occur in the area of the falx cerebri, on the floor of the middle cranial fossa and on the upper and lower surfaces of the tentorium cerebelli. On the top surface of the tentorium these bridging veins were found to be nearly 12 mm long on average. Commonly these bridging veins turn into so-called sinus tentorii, that we interpret as blood spaces within the tentorium cerebelli. Their number on the upper surface of the tentorium is listed in Fig. 15. These sinuses mostly do not run straight but curve to their confluence sites. As sinus tentorii dextri laterales their width averages 3.8 (2–6) mm and their height 1–2 mm. It is likely that by moving the endoscope forward or backward, transcisternal or bridging veins can easily be damaged.

References

1. Bisaria KK (1985) The superficial sylvian vein in humans: With special reference to its termination. Anat Rec 212: 319–325
2. Burman WE (1931) Myeloscopy or the direct visualization of spinal canal and its contents. J Bone Jt Surg 13: 695–696
3. Dandy WE (1922) Cerebral ventriculoscopy. Bull John Hopkins Hosp 33: 189
4. Deymann-Bühler B (1984) Maße des Mesencephalons und des Ventriculus IV. Med Diss, Würzburg
5. Fay T, Grant FC (1923) Ventriculoscopy and intraventricular photography in internal hydrocephalus. Report of case. JAMA 80: 461–463
6. Fukushima T (1978) Endoscopy of Meckel's cave, cisterna

magna, and cerebellopontine angle. J Neurosurg 48: 302–306

7. Hussein S, Woischneck D (1990) Topographie des Atrium ventriculi und ihre mikrochirurgische Bedeutung. Neurochirurgia 33: 8–10

8. Key A, Retzius G (1875) Studien in der Anatomie des Nervensystems und des Bindegewebes. Bd 1. Samson & Wallin, Stockholm

9. Koelliker A (1850) Mikroskopische Anatomie. Vol II, cited from Yaşargil et al, 1984

10. Krauss J (1987) Messungen zur cranio-cerebralen Topographie. Med Diss, Würzburg

11. Lang J (1973) Die äußeren Liquorräume des Gehirns. Acta Anat 86: 267–299

12. Lang J (1979) Praktische Anatomie. Ein Lehr- und Hilfsbuch der anatomischen Grundlagen ärztlichen Handelns. Begr v T von Lanz, W Wachsmuth. Fortgef u hrsg von J Lang, W Wachsmuth. Teil 1: Bd 1 Kopf. Teil B Gehirn- und Augenschädel, von J Lang in Zsarb mit K-A Bushe, W Buschmann, D Linnert. Springer, Berlin Heidelberg New York

13. Lang J (1985) Lanz/ Wachsmuth. Praktische Anatomie. Bd I/1 Kopf. Teil A: Übergeordnete Systeme, von J Lang. In Zsarb mit H-P Jensen, F Schröder. Springer, Berlin Heidelberg New York

14. Lang J, Debes K (1977) Über Kleinhirnvenen und Drainagegebiete des Cerebellum. Verh Anat Ges 71: 719–723

15. Lang J, Emminger A (1963) Über die Textur des Ligamentum denticulatum und der Pia mater spinalis. Z Anat Entw-Gesch 123: 505–522

16. Lang J, Köth R, Reiss G (1981) Über die Bildung, die Zuflüsse und den Verlauf der V. basalis und der V. cerebri interna. Anat Anz 150: 385–423

17. Lang J, Nadjmi M (1973) Anatomie und Radiologie der zerebralen Subarachnoidealräume. Rö-Bl 26: 107–112

18. Lang J, Schaffrath H, Fischer G (1987) Weitere Befunde zu den Rami diencephalici. Neurochirurgia 30: 103–107

19. Lang J, Stefanec P, Breitenbach W (1983) Über Form und Maße des Ventriculus tertius, von Sehbahnteilen und des N. oculomotorius. Neurochirurgia 26: 1–5

20. Lang J jr, Ohmachi N, Lang J Sen (1991) Anatomical Landmarks of the rhomboid fossa, its length and its width. Acta Neurochir (Wien) 113: 84–90

21. Liliequist B (1959) The subarachnoid cisterns. An anatomic and roentgenologic study. Acta Radiol (Stockh) [Suppl] 185: 1–108

22. Lindgren E, DiChiro G (1953) The roentgenologic appearance of the aqueduct of Sylvius. Acta Radiol 39: 117, cited from Liliequist B 1959

23. Lindstrom PA, Brizzee KR (1962) Relief of intractable vomiting from surgical lesions in the area postrema. J Neurosurg 19: 228–236

24. Magendie F (1822) Lecons sur les fonctions et les maladies du système nerveux. Ebrard, Paris

25. Marin OS, Angevine JB, Locke S (1962) Topographical organization of the lateral segment of the basis pedunculi in man. J Comp Neurol (Philadelphia) 118: 165–183

26. Mixter WJ (1923) Ventriculoscopy and puncture of the floor of the third ventricle, preliminary report of a case. Boston Med Surg J 188: 277–278

27. Nomina Anatomica (1989) Sixth Edition. Authorized by the Twelfth International Congress of Anatomists in London, 1985, together with Nomina Histologica. Third Edition. Nomina Embryologica. Third Edition. Revised by Subcommittees of the International Anatomical Nomenclature Committee. Churchill Livingstone, Edinburgh London Melbourne New York

28. Pool JL (1938) Direct visualization of dorsal nerve roots of cauda equina by means of a myeloscope. Arch Neurol Psychiat 39: 1308–1312

29. Peter K (1938) Die Nase des Kindes. In: Peter K, Wetzel G, Heiderich F (Hrsg) Handbuch der Anatomie des Kindes. Bd II. JF Bergmann, München

30. Prott W (1974) Untersuchungen zur Endoskopierbarkeit des inneren Gehörganges und des Kleinhirnbrückenwinkels auf otochirurgischen Zugangswegen (Meatocisternoskopie). Habilitationsschrift, Würzburg

31. Putnam TJ (1934) Treatment of hydrocephalus by endoscopic coagulation of the choroid plexus. Description of a new instrument and preliminary report of results. N Engl J Med 210: 1373–1376

32. Renella RR (1989) Microsurgery of the temporo-medial region. Springer, Wien New York

33. Robertson EG (1941) Encephalography. MacMillan & Co, Ltd, Melbourne

34. Samii M, Draf W (1989) Surgery of the skull base. An interdisciplinary approach. With a chapter on anatomy by J Lang. Springer, Berlin Heidelberg New York London Paris Tokyo Hong Kong

35. Scarff JE (1935) Third ventriculoscopy as the rational treatment of obstructive hydrocephalus. J Pediatr 6: 870–871

36. Schaltenbrand G (1955) Plexus und Meningen. In: Bargmann W (Hrsg) Handbuch der mikroskopischen Anatomie des Menschen. Bd 4, Teil 2. Springer, Berlin Göttingen Heidelberg

37. Schubert M (1976) Praktisch-anatomische Befunde in der Fossa cranii posterior. Med Diss, Würzburg

38. Stern EL (1936) Spinascope: new instrument for visualizing the spinal canal and its contents. Med Rec 143: 31–32

39. Sunderland S (1958) The tentorial notch and complications produced by herniations of the brain through that aperture. Br J Surg 45: 422–438

40. Taveras JM, Wood EH (1964) Diagnostic neuroradiology. Williams and Wilkins, Baltimore

41. Torkildsen A (1933/34) The gross anatomy of the lateral ventricles. J Anat 68: 480–491

42. Turnbull IM, Drake CG (1966) Membraneous occlusion of the aqueduct of Sylvius. J Neurosurg 24: 24–33

43. Umbach W (1960) Die operative Behandlung des Hydrocephalus: Plexusextirpation und -koagulation. In: Olivecrona H, Tönnis (eds) Handbuch der Neurochirurgie, Klinik und Behandlung I. Springer, Berlin Heidelberg New York, pp 657–661

44. Yaşargil MG, Smith RD, Young PH, Teddy PJ (1984) Microsurgical anatomy of the basal cisterns and vessels of the brain, diagnostic studies, general operative techniques and pathological considerations of the intracranial aneurysms. Thieme, Stuttgart New York

Correspondence: Prof. Dr. med. J. Lang, Department of Anatomy, University of Würzburg, Koellikerstrasse 6, D-W-8700 Würzburg, Federal Republic of Germany.

Acta Neurochirurgica, Suppl. 54, 11–25 (1992)
© by Springer-Verlag 1992

Endofiberscopic Intracranial Stereotopography and Endofiberscopic Neurosurgery

V.B. Karakhan

Department of Neurology and Neurosurgery, Moscow Medical Stomatological Institute, USSR

Summary

The possibilities of endoscopic operative procedures in preformed and artificial intracranial and intracerebral cavities are described. Endoscopic investigations were carried out on 166 cadavers; death was due to extracranial pathology and severe brain injury.

132 patients with intracranial lesions were operated on endoscopically.

An extended description of endofiberscopic intracranial topographical relationships and endofiberscopic neurosurgical approaches is given. Endoscopic stereotopography of more than 130 intracranial structures is studied.

Keywords: Intracranial endoscopy; stereotopography; endofiberscopic neurosurgical approaches.

Abbreviations

-a-	ala parva
-e-	epiphysis et comissura habenularum
f	lobus frontalis
h	hippocampus
t	lobus temporalis
aq	aqueductus cerebri
ca	comissura anterior
cm	corpus mamillare
cp	comissura posterior
fi	foramen interventriculare
adh	adhesio interthalamica
crb	cerebellum
D	Dura
M	Meatus acusticus internus
P	Posterior wall of the petrous temporal bone
T	Tentorium cerebelli
ACS	arteria cerebelli superior
ACIA	arteria cerebelli inferior anterior
ACIP	arteria cerebelli inferior posterior
AB	arteria basilaris
AV	arteria vertebralis
ACI	arteria carotis interna
ACA	arteria cerebri anterior
ACM	arteria cerebri media
ACP	arteria cerebri posterior
ACoP	arteria communicans posterior
AChA	arteria chorioidea anterior
II–XII	cranial nerves
II$_t$	tractus opticus
C 1	radices spinales I

Introduction

Modern endoscopic techniques, which allow one to observe and manipulate beyond the direct vision under optic magnification open wide opportunities for its usage in endocranioscopy. There are three reasons:

Trauma free observation of deep structures, which is the basis for assessment of the normal relations in microsurgical neuroanatomy and features of intracranial stereotopography;

The observation and interventions beyond the opening projection provide considerably less trauma in the surgical approach.

The magnified picture of structures, being observed allows one to muster microsurgical methods.

Endoscopic neurosurgical interventions need an extended investigation and description of the topographical relationships of intracranial and intracerebral structures, which is the main topic of this work.

Material and Methods

1. Endofiberscopes

It is known that endoscopic observation is effective when there is a gap between lens and structure within the limits of the cavity.

Multilevel intracranial spaces have complex and tortuous outlines. Therefore specific stereotopographic approaches have been used in order to reach various intracranial areas and structures indirectly, using parabolic traces which are oriented on certain landmarks.

Table 1. *Instrumentation for Endocranioscopy*

1. Endofiberscopes ("Olympus", Japan)

Name	Outer diameter	Channel
ENF – P	3.7 mm	–
CHF – B	5.9 mm	2 mm
BF – B3R	5.9 mm	2 mm
CHF – B3R	6.7 mm	2.6 mm

2. Endoscopic instruments (flexible pivot)

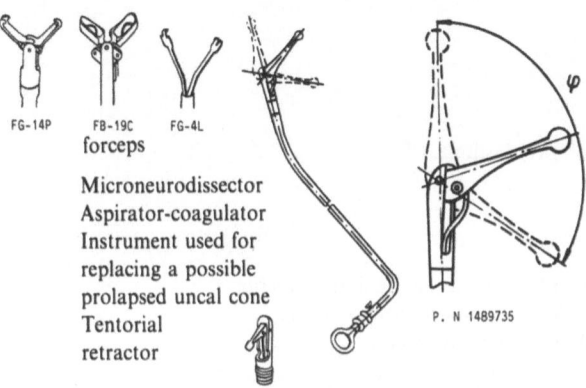

FG-14P   FB-19C   FG-4L

forceps

Microneurodissector
Aspirator-coagulator
Instrument used for
replacing a possible
prolapsed uncal cone
Tentorial
retractor

P. N 1489735

Table 2. *Basic Methods of Endoscopic Removal of Extracerebral Haematomas*

Actions	Tools
1. Clot mobilization and aspiration from the haematoma central mass	Draining tube Endoscopic dissector
2. Clot mobilization and washing out from the haematoma borders	Channel of endoscope Endoscopic dissector
3. Aspiration and extraction of large clots near the burr hole	Draining tube Endoscopic forceps FG-4L
4. Draining of liquid mass, delimited by clots, brain convolutions or haematoma membranes	Tip of endofiberscope Endoscopic dissector Endoscopic forceps
5. Separation of inner membrane of chronic subdural haematoma from arachnoidal surface	Tip of endofiberscope Endoscopic dissector

Endoscopic removal of adjacent intracerebral haematomas

* Clot aspiration and washing out into available cortical opening without additional brain perforation	Channel of endoscope Endoscopic forceps FG-4L

For this reason flexible instruments are necessary.
The main specifications of the endofiberscopes, which were used for our studies, are listed in Table 1.1.

2. Endoscopic Instruments

Special instruments were developed for application in endoscopic interventions. They are described in Table 1.2.

3. Endoscopic Procedures

Topographical studies: Intracranial endoscopic investigations in 166 cadavers; in whom death was due to extracranial and severe head injury, were carried out and the topographic-anatomical relationships are described.

Operative treatment: 132 patients were operated on endoscopically, 119 of these patients suffered from traumatic intracranial haemorrhages, the remaining 13 patients had ventricular tumours, cysts or neurovascular compression of the cerebello-pontine angle.

Haematoma evacuation: The methods of endoscopic removal of intracranial haematomas are listed in Table 2.

Dense blood-clots are mobilized by dissector. Dealing with liquid haematomas and hydromas, endoscopy allows one to perforate the restricted blood filled cavities or make an arachnoid opening into the cisterns.

Endoscopic removal of intracerebral haematomas adjacent to the subdural space is performed through the available cortical opening without additional brain perforation, preserving also the bridging vessels.

These methods allow one to remove also bilateral foci.

Acute tentorial pressure cone: The endoscopic signs of acute temporal pressure cone have been investigated by the subtentorial retromastoid approach. A model with a supratentorial blowing balloon in cadavers is used, to gain information which was compared with the results observed in patients.

Severance of neuro-vascular contacts: The methods of endoscopic severance of neurovascular contacts in the cerebello-pontine angle with the help of endoscopic dissectors are developed. For reliable neurovascular disconnection microneuroprotectors have been developed and used. Neuroprotector (Fig. 13 c, e) consists of two long semicircular prongs and two short ones. The device is hollow, with elastic walls. Forceps grasping the short prongs leads to opening of the long prongs. After trigeminal root-release from arterial loop (Fig. 13 a) the open protector approaches the nerve (Fig. 13 b, c) with subsequent nerve embracing by long prongs. The arterial loop is inserted between the short prongs. Elastic clip-shaped protectors for application upon an artery have been developed also (Fig. 14 d). Identification of compressing artery has been performed by segmental endoscopic examination of the vessel course.

Results

1. Endofiberscopic Intracranial Stereotopography

a) The Subdural Endoscopic Stereotopography (Figs. 2 and 3)

Through a lateral temporo-parietal opening, one inch in diameter, one can observe the largest part of the convexity and laterobasal hemispheric surface. It is important to control simultaneously in the endoscopic visual field the dural and cortical surfaces in order to avoid damage. During the course of endoscopic examination the bridging vessels are seen (Fig. 2 e). The bridging veins at the place of attachment to dura have a red halo which corresponds to arachnoidal sus-

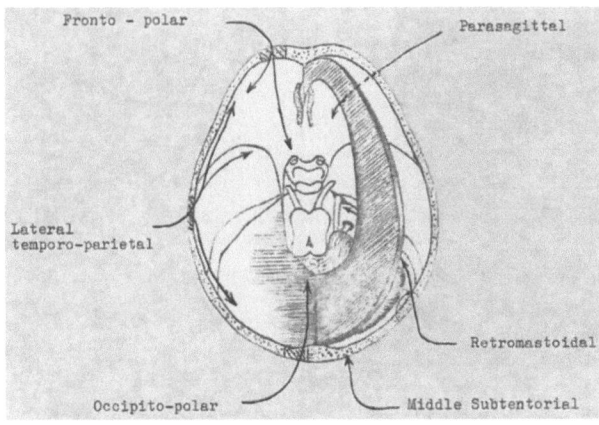

Fig. 1. Topography of main intracranial endofiberscopic approaches

pending adhesions (Fig. 2 f). The bending of the distal endoscopic segment allows one to avoid injury of these bridging structures.

For precise space-orientation in convexity subdural areas, which contain a cortical and dural surface only (Fig. 2 a), the phenomenon of intracranial endoscopic transillumination has been investigated and used. This is a bright red spot appearing on the scalp according to the projection of the endofiberscopic tip. In this way it is found that for reliable endoscopic lateral subdural examination only 6 directions of tube insertion are necessary. Initial direction is to the sphenoidal wing. Depending on the site of observation it has a wedge (at the level of the wing edge) or crescent shape (Figs. 2 b and 3). The sphenoidal wing is a key-directing structure for endoscopic subdural orientation. It distinguishes frontal and temporal lobe, middle and anterior fossa. A gap between frontal lobe and wing edge leads to the anterior fossa. The orbital view is revealed in lateral areas and optic nerve-in medial ones (Fig. 2 c, d). One can count the orbital convolutions. In front of them the bright surface of falx cerebri is visualized.

By fronto-polar observation the optic nerves, olfactory tract, and the convolution of the gyrus rectus are revealed. Through an occipito-polar opening the medical surface of occipital lobe, the splenium of the corpus callosum, lamina quadrigemina at the bottom and the interhemispheric fissure, the sinus sagittalis superior with draining veins at the top are revealed.

b) The Endoscopic Cisternal Topography (Fig. 4)

The subtentorial retromastoid approach allows one to investigate the endoscopic stereotopography of the basal arachnoid cisterns (Fig. 4). It is important first of all for selective incision of cisternal walls and the determination of vessel courses and their contacts. Walls of cisterns are stretched between cranial nerves. Endoscopic observation and microdissection of their wall without structural shift allow one to outline: the lateral pontine cistern (arachnoid membrane is stretched between trigeminal and facial nerves; one can see the subdural mouth of the petrosal vein, Fig. 4 a) middle pontine cistern (between trigeminal and abducent nerves; Fig. 4 b), ambient cistern (between trigeminal and trochlear nerves; Fig. 4 c), crural and interpeduncular cisterns (Fig. 4 d).

It allows one to detect cisternal haemorrhages for endoscopic draining of: lateral pontine, middle pontine, ambient, and interpeduncular cisterns.

c) The Endoscopic Neural Stereotopography (Fig. 5)

After microdissection of the cisternal walls the neural roots are revealed. Observation of the caudal nerve group allows one to determine N. IX, X, XI (Fig. 5 f). Through the gap between the glossopharyngeal and vagus nerves one can see the hypoglossal nerve, olive, margo foraminis occipitalis and cisterna magna. The skeletotopic finding for the facial and acoustic nerves is the bony ridge above the nerves-the portal (Fig. 5 b)-unlike the trigeminal nerve.

d) Petrosal Veins, Endoscopic Topography (Fig. 6)

These veins cross the approach to the cerebellopontine angle. It is interesting that the vein mouth detected by halo, may be shifted from the superior petrosal sinus projection. Petrosal veins endofiberscopic assessment is described elsewhere[1].

e) The Endoscopic Arterial Topography (Fig. 7)

The endoscopic stereotopography of the arterial circle of Willis is investigated through the same subtentorial retromastoid access (Fig. 7). One can distinguish: a.carotis interna, a.cerebri media, a. cerebri anterior, a.communicans posterior, a.choroidea anterior. The vertebro-basilar system is visualized: a.basilaris, a.cerebri posterior dextra et sinistra, (P1 and P2 segments) a.communicans posterior, a.cerebelli superior, hippocampus, a.cerebelli inferior anterior cum arteria labyrinthi, a.cerebelli inferior posterior, passing from vertebral artery near the olive and N.XII.

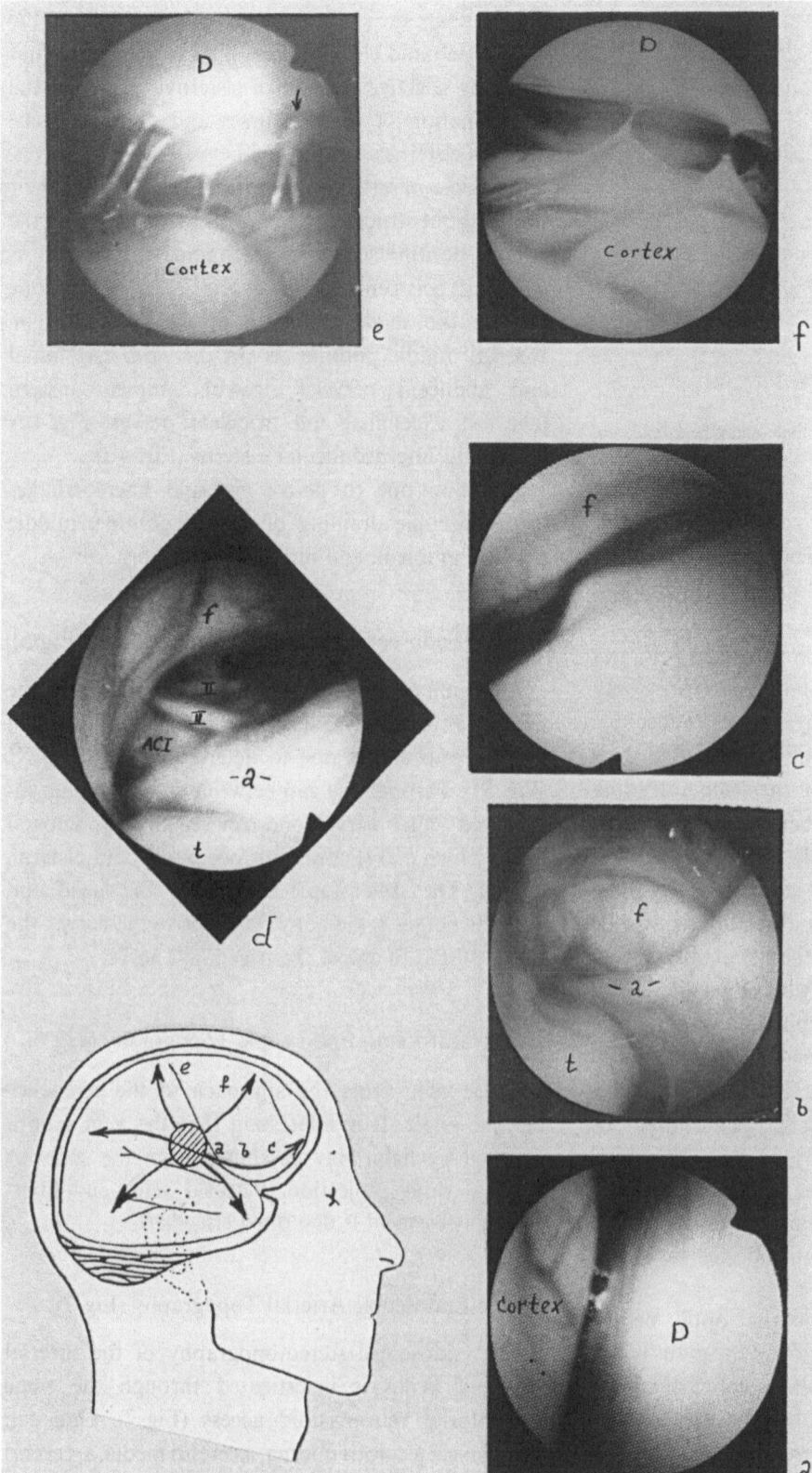

Fig. 2. Subdural endoscopic stereo-topography. a) Temporoparietal area, vertical orientation of sub-dural space; limited by cortical and dural surfaces, b) small wing, dividing frontal and temporal lobe, c) view of the anterior fossa, horizontal orientation of subdural space, d) medial part of lesser wing, of sphenoid the optic nerves are out-lined, e) multiple bridging veins between parasagittal cortex and dural surface with halo at the level of vein opening (arrow), f) arach-noidal connections, suspending inter-convolutional arteries

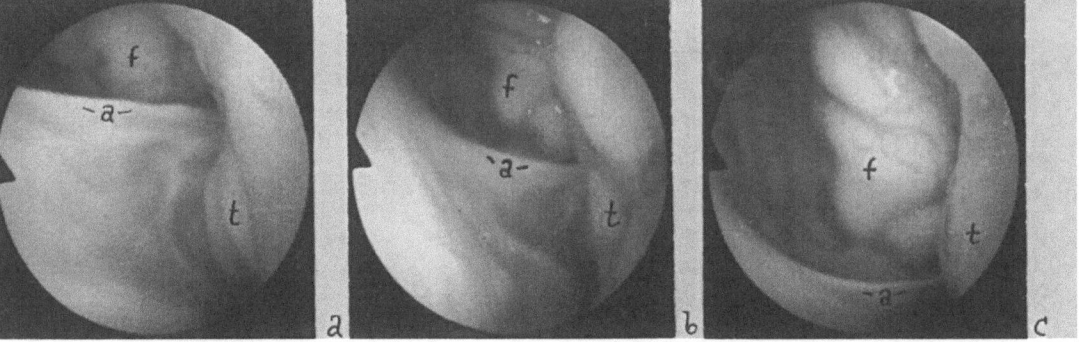

Fig. 3. Peculiarities of endoscopic view of the small wing (on the left). a–c) "Disappearing moon" in serial pictures, when the endoscope passes from middle to anterior fossa

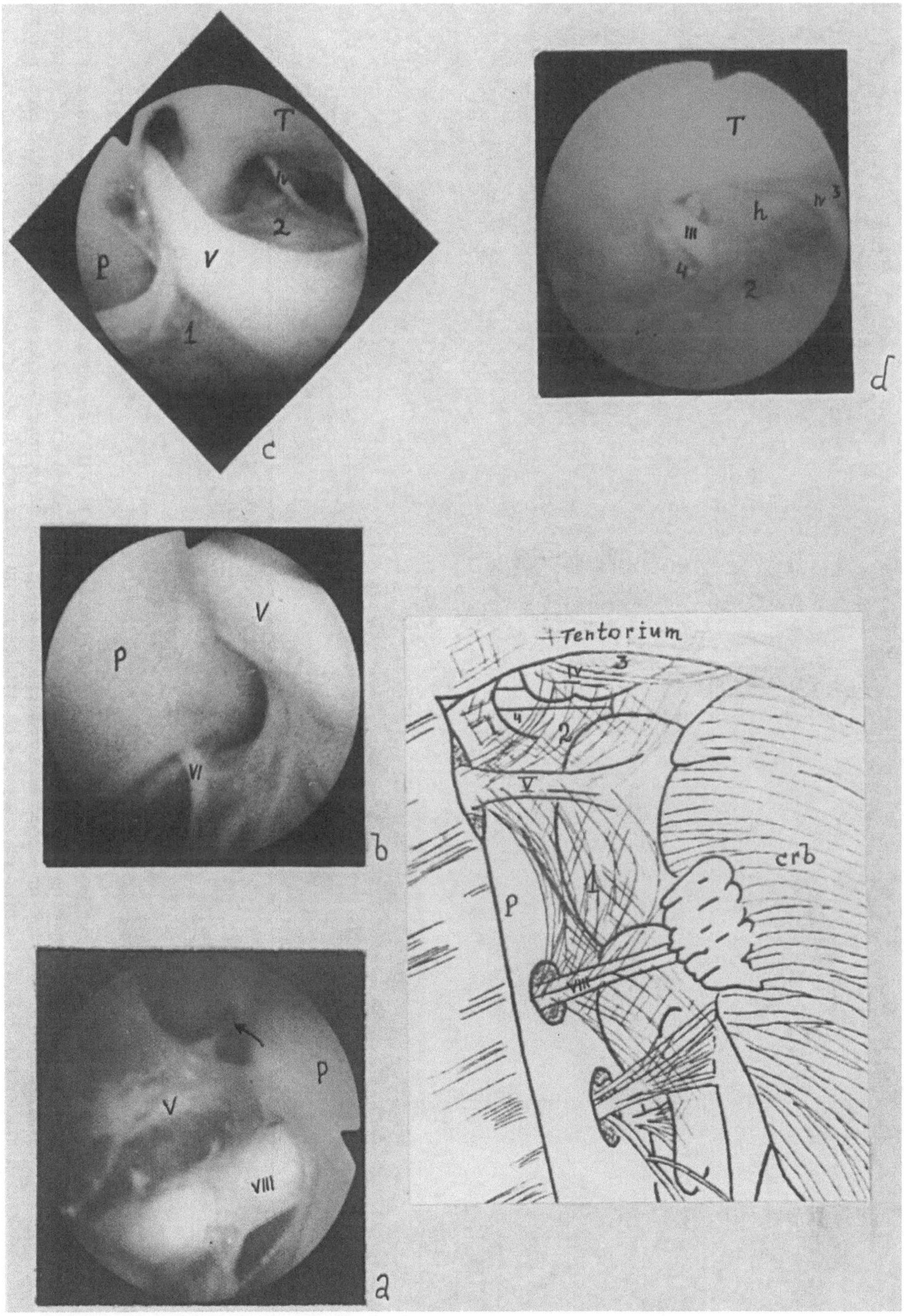

Fig. 4. Endoscopic cisternotopography (retromastoid subtentorial approach. a) Lateral pontine cistern, subdural site of petrosal vein terminal (arrow), b) middle pontine cistern, c) endoscopic assessment of the cisternal relationship. (1) Lateral pontine cistern, (2) ambient, (3) crural, (4) interpeduncular cistern

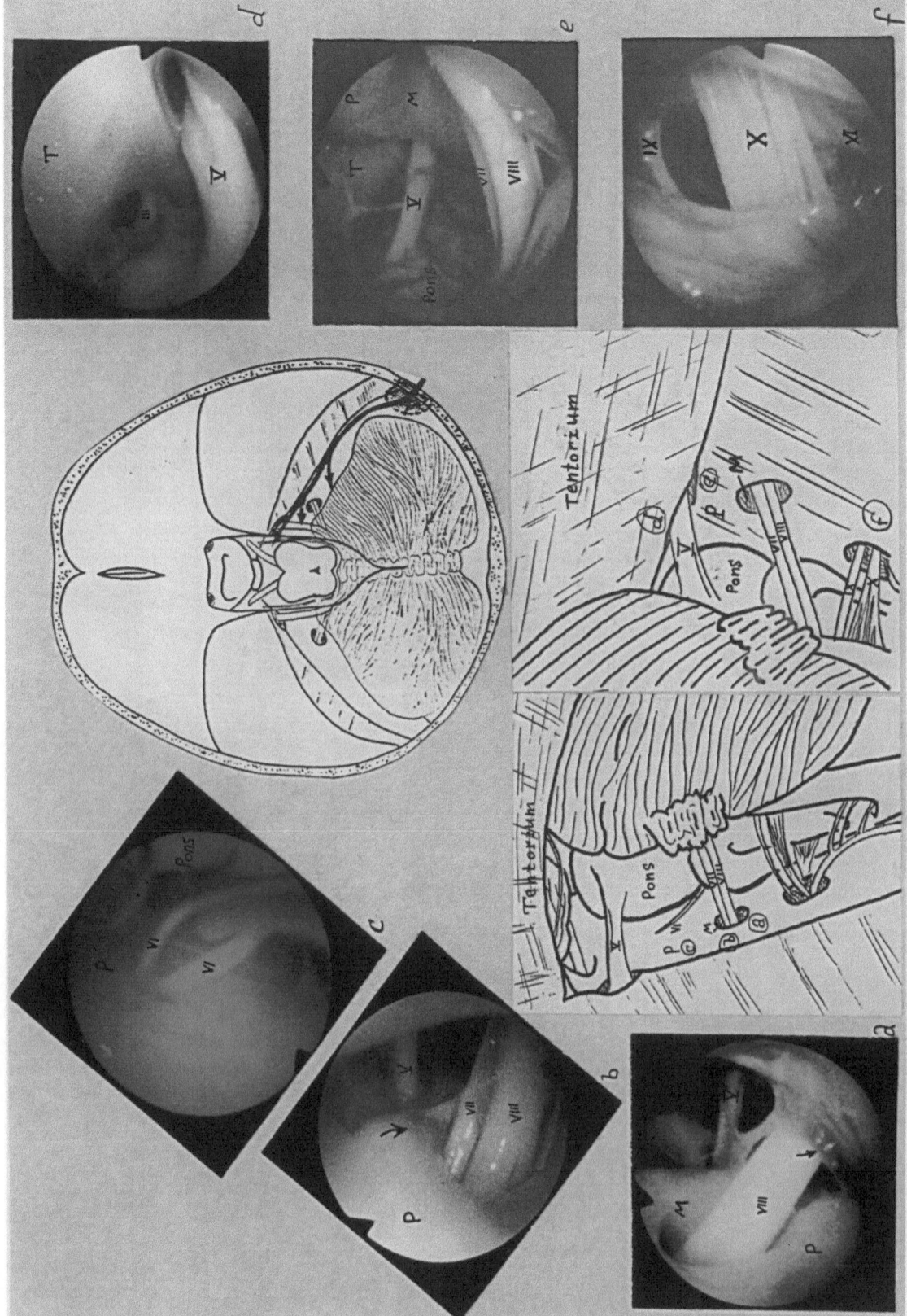

Fig. 5. Endoscopic neurotopography in the cerebellopontine angle. a) N. VIII, internal acoustic meatus, cross-contact of nerve with ACIA loop (arrow), b) divided visualization of facial and acoustic nerves, bone apophysis passes between them, c) abducent nerve split into two roots, ACIA passes between them, d) band-shaped rotated trigeminal root (endoscpic tip is over the nerve), e) relationship of N. V. (cylindric-shaped), N. VII, N. VIII and pons, pyramis posterior wall (endoscopic tip is at the level of N. VIII), f) bulbar nerve group. The sketches present traces and site of observation

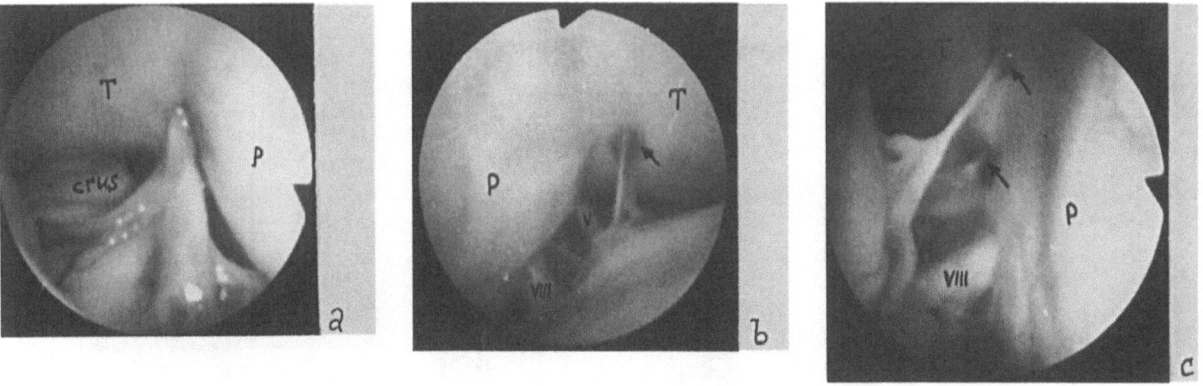

Fig. 6. Endoscopy of petrosal veins (retromastoideal approach). a) One wide vein and its branches, b) lambda-shaped conjunction of branches with a shift of the mouth (arrow) to the tentorium, c) two venous tubes (medial and lateral from internal meatus) and their mouths (arrows) projectioning the superior petrosal sinus trace

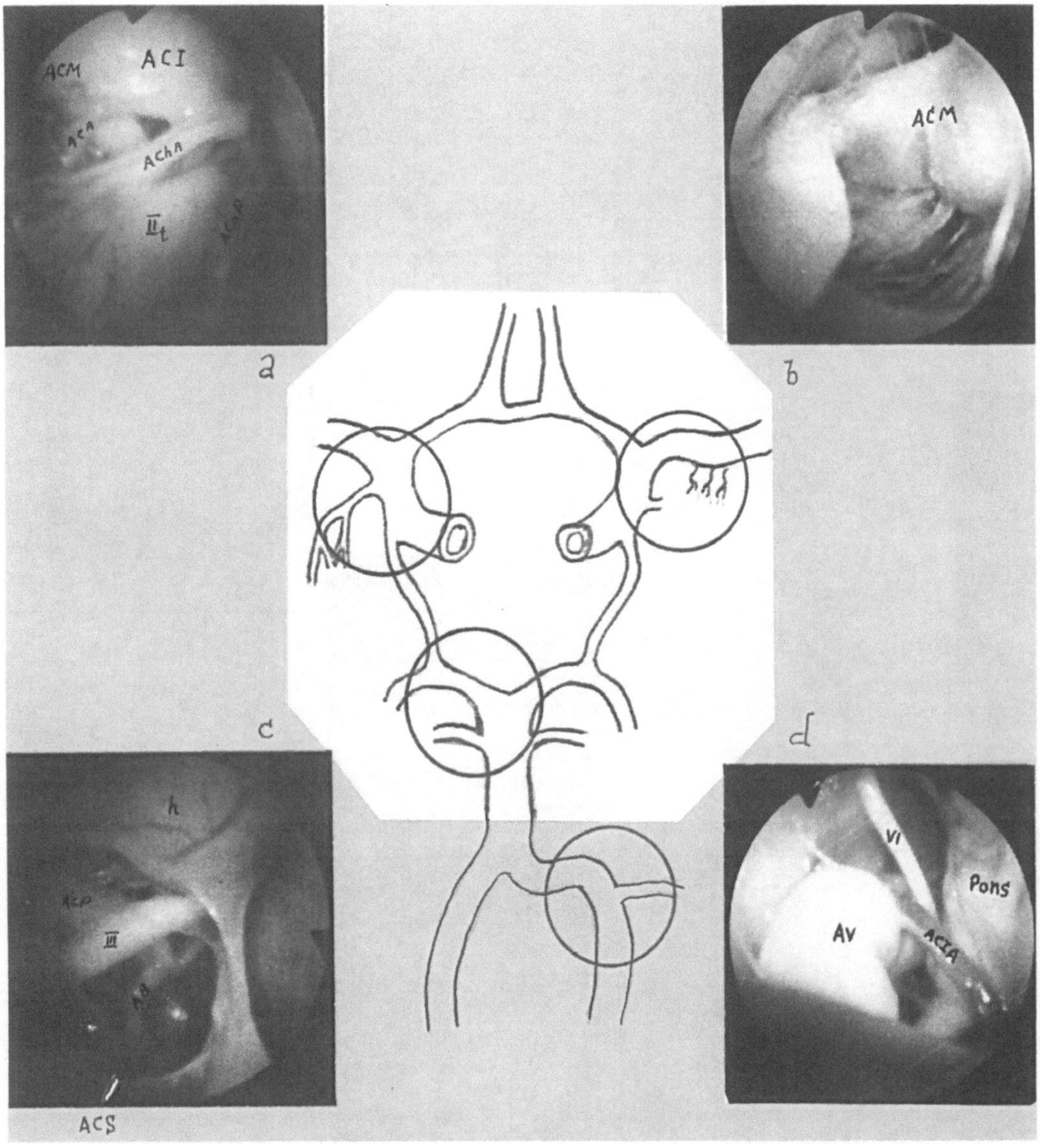

Fig. 7. Endoscopic stereotopography of the circle of Willis and branches (retromastoid subtentorial approach). In the sketch all sites of observation are outlined

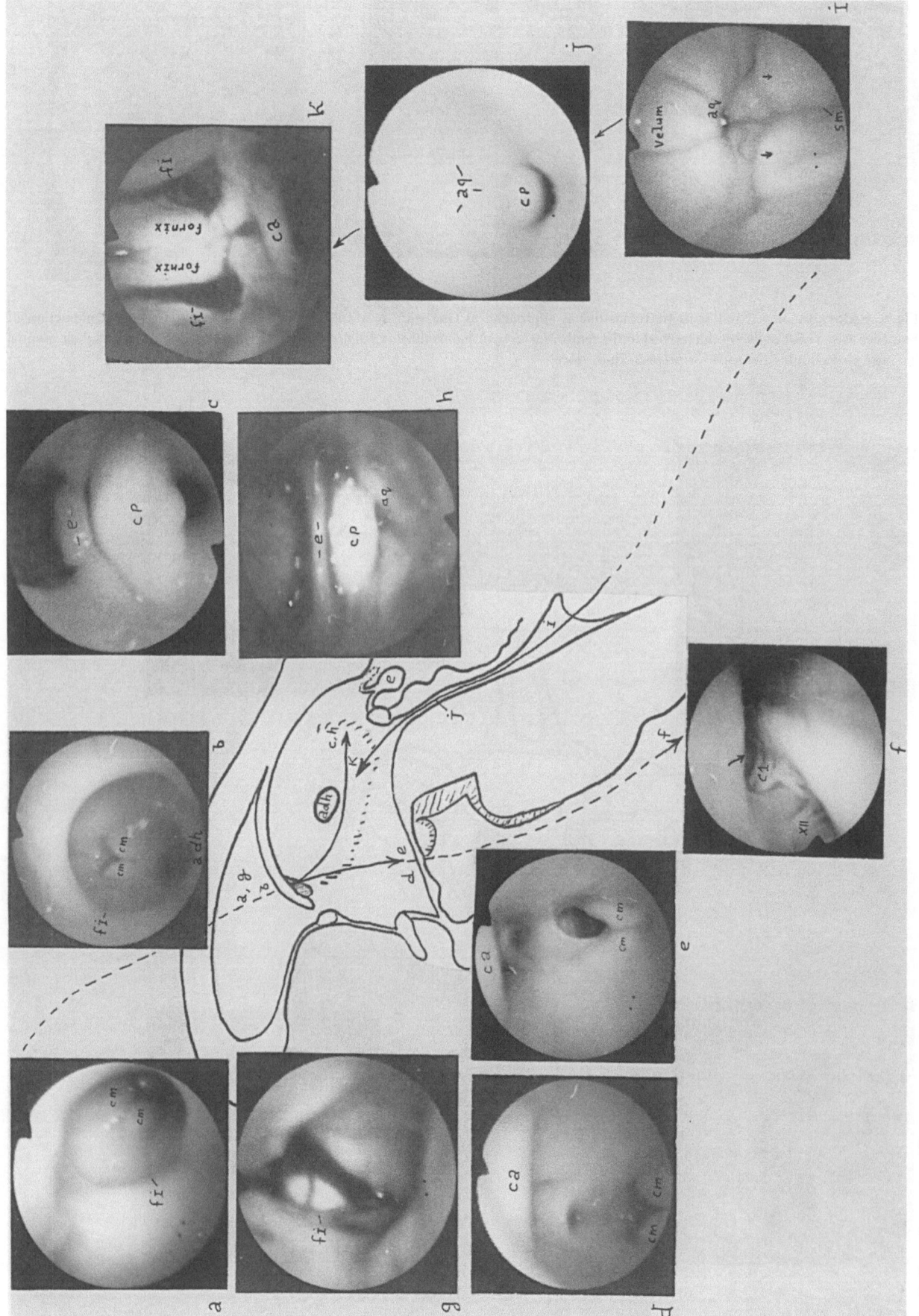

Fig. 8 (legend see page 19)

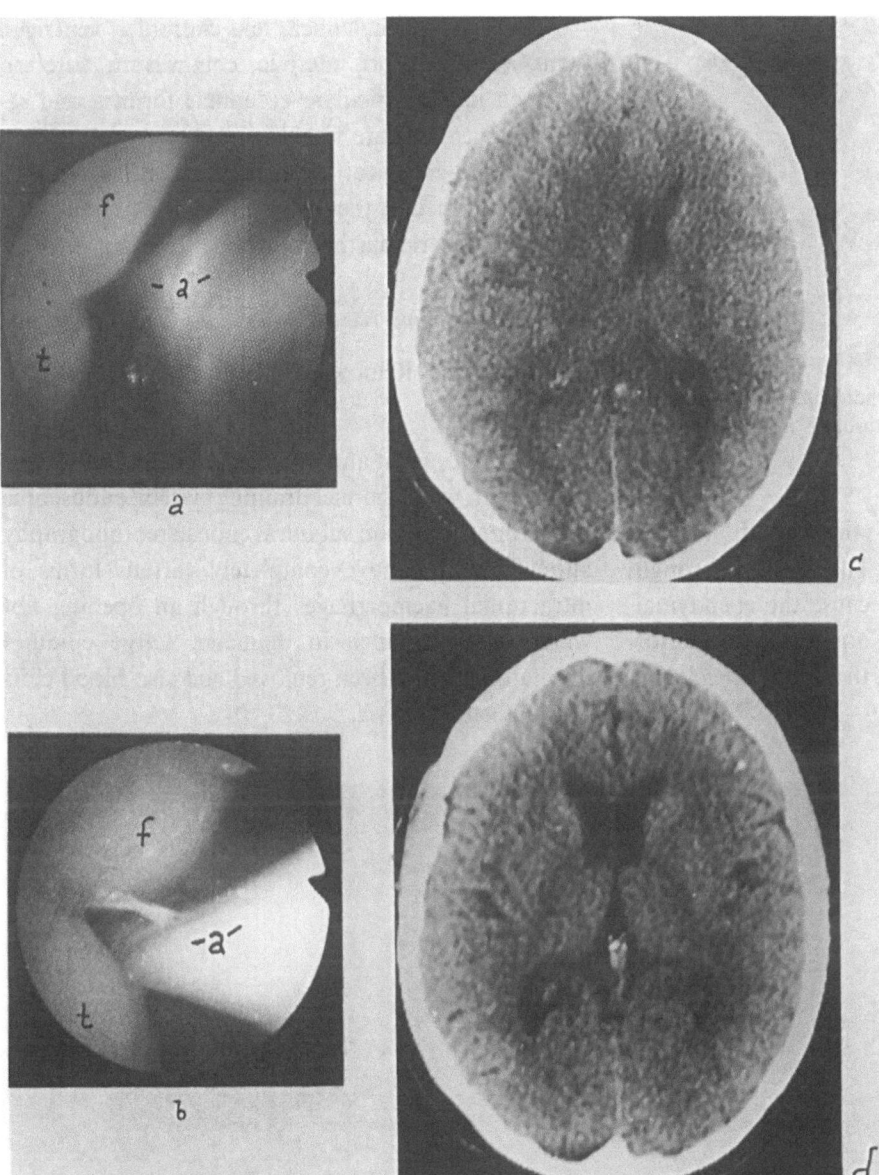

Fig. 9. Endoscopic and computer-tomographic pictures before (a, c) and after (b, d) withdrawal of a wide-spread isodense subdural haematoma

f) Endoscopic Ventricular Topography (Fig. 8)

Through the anterior horn approach the walls of the lateral ventricle with the choroid plexus are revealed, the interventricular foramen, posterior part of the third ventricle with aqueduct entry, posterior commissure, pineal and suprapineal recesses and the floor of the third venetricle with infundibular and chiasmatic recesses are outlined (Fig. 8 a–d).

By means of the endoscopic tip the posterior wall of the tuber cinereum (between the corpora mamillaria,

Fig. 8 e) is incised. The middle pontine cistern is penetrated (depth about 6 mm) between dorsum sellae and the basilar bifurcation. This way leads to the edge of the foramen magnum, N.XII and the first spinal roots (Fig. 8 f).

The ascendent endoscopic approach to ventricular system is investigated (Fig. 8 i–k). Through the middle occipital opening the thin endoscope (smaller than 4 mm in diameter) is inserted between lingula and obex. In this way the fourth ventricle is observed.

Colliculi faciales, calamus scriptorius, vela medul-

Fig. 8. Endofiberscopic ventriculotopography (a–h) through anterior horn of lateral ventricle, (i–k) through fourth ventricle with recording of ventricular haemorrhages. a, b) Foramen interventriculare and view of third ventricle floor, c) posterior part of third ventricle, d, e) topography of third ventricle before and after premammillary perforation, f) transcisternal access of foramen magnum (arrow) area, g) periforaminal haemorrhage, h) blood in the pineal, subpineal recesses and in the aqueduct, i) topography of the fourth ventricle (colliculi faciales are marked by arrows), j) intraaqueductal approach, k) topography of the anterior part of the third ventricle with periforaminal haemorrhages

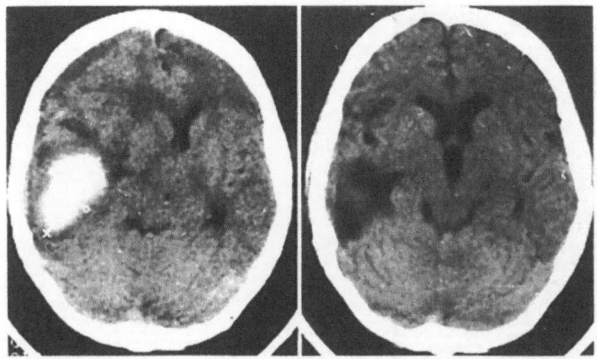

Fig. 10. CT pictures before and after endoscopic evacuation of an adjacent intracerebral haematoma through available cortical perforation opening

lares, striae medullares and the mouth of the aquaeduct are outlined (Fig. 8 i). Tube passing through the aquaeduct is atraumatic because the ependymal folds are smoothed out. The aquaeduct, posterior commissure (Fig. 8 j) and at last, the third ventricle are visible (Fig. 8 k).

Adhaesio interthalamica, tela choroidea ventriculi tertii, venae cerebri internae, commissura anterior, foramen interventriculare columnae fornicis and recessus chiasmatis are visualised.

The topographical investigations allow one to determine precisely the site of ventricular tumours and spread of ventricular haemorrhages (Fig. 8 g, h, k).

2. Endofiberscopic Neurosurgery

a) Endoscopic Removal of Intracranial Traumatic Haemorrhages

The development and usage of a coaxial endoscopic nozzle with wide-channel draining system, endoscopic dissector and data on subdural endostereotopography allow one to remove completely various forms of intracranial haemorrhages through an opening not wider than one inch in diameter. Large epidural haematomas have been removed and also blood clots of the posterior fossa.

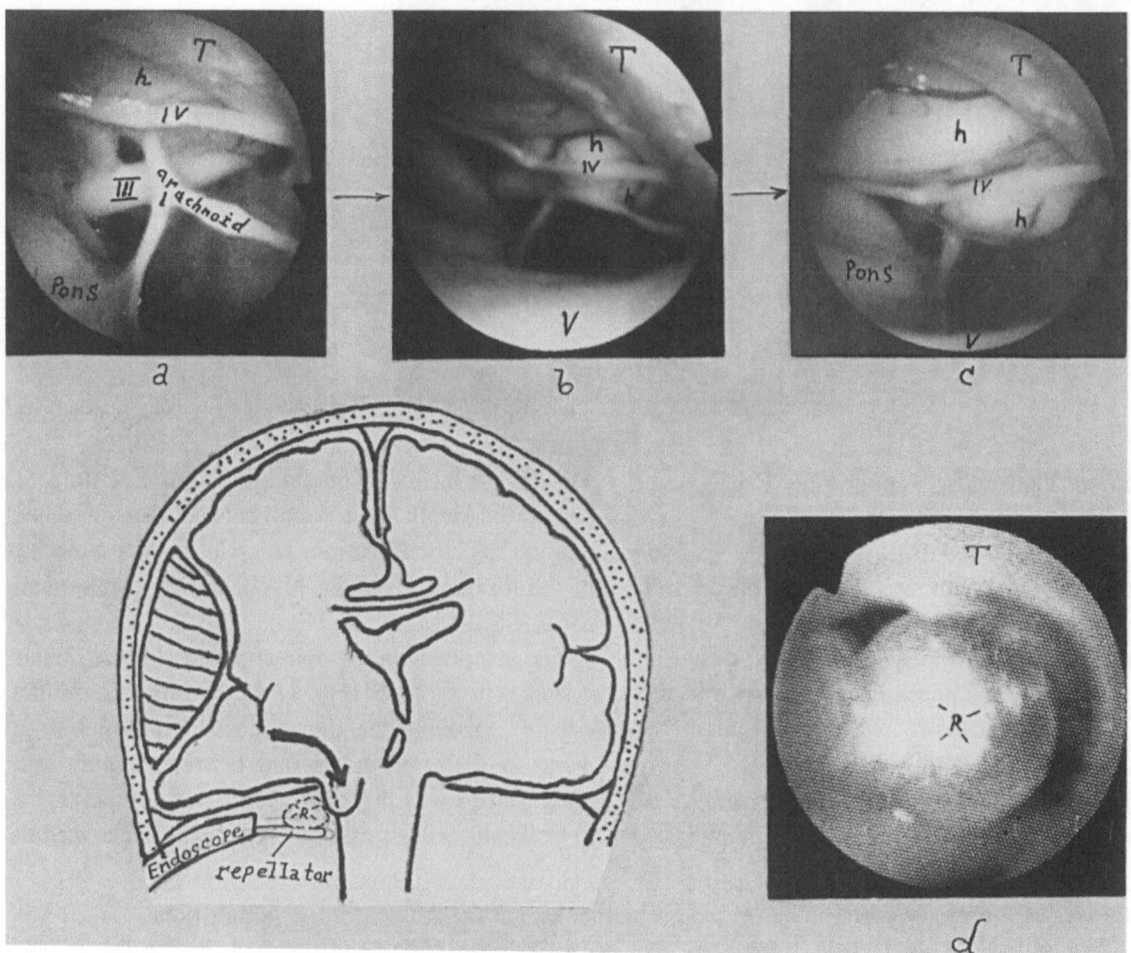

Fig. 11. Stages of acute tentorial pressure cone formation. (a–c) Right retromastoid approach and technique of endoscopic replacement of herniated hippocampus (d), a) normal topography of right tentorial edge area, b, c) growing hippocampal herniation with dislocation of adjacent structures. A sketch shows the balloon in action as it is gradually blown up

Table 3. *Trephine Endoscopic Surgery of Traumatic Intracranial Haemorrhages*

	Surgical cases	Volume	Relapse	Dead
Epidural haematomas (posterior fossa)	18 (2)	50–200 ml	1	1
Subdural haematomas	69	60–200 ml	2	10
Subdural hygromas	10	50–120 ml		3
Intracerebral adjacent haematomas	7	40–80 ml	–	1
Intraventricular haemorrhage	2	30–40 ml	–	1
Brain laceration	3	30–60 ml	–	
Total	109		3	16

Bilateral – 7,   Decompressive trepanation – 10
Acute, subacute forms – 76   (death rate – 21%)
Operated in several stage – 63   (death rate – 23%)

The visualisation of the attachment of the dura to the bone at the margin of the haematoma is a criterion for total clot removal.

As an example for successful treatment endoscopic and computertomographic pictures before and after removal of widespread subdural clots (Fig. 9 a–d) and of intracerebral haematoma (Fig. 10) are presented.

The overall results of endoscopic haematoma evacuation are described in Table 3.

b) Endoscopic Assessment and Correction of Acute Tentorial Pressure Cone

The serial pictures of a growing right tentorial cone, caused by supratentorial pressure are taken in the balloon-model described (Fig. 11 a–c).

General forms of tentorial pressure cone, depending on the site of compression are: anterior drop-shaped hernia, or a posterior semicircle form.

Modelling results are like the ones, observed in patients, who died due to severe brain injury. The signs of nerve-, brain stem-, vessel- and cisternal wall dislocation have been investigated by continuous one- or simultaneous double-sided observations with the help of two endoscopes. The results are described elsewhere[2].

The replacement of herniated convolutions by means of a suitably placed microballoon has been experimentally developed (Fig. 11 d) and the method of endoscopic ventriculo-cisternal drainage was performed (Fig. 8 e).

c) Endoscopic Correction of Neurovascular Contacts in Cerebello-Pontine Angle

An endofiberscopic assessment of compressing neural contacts has been made (Figs. 12–17).

The general forms of artery/trigeminal root contacts are:

1. the upper lateral prepontine one with a superior cerebellar artery loop (Fig. 12 a);
2. the upper medial one with the same artery (Fig. 13 a);
3. the contact of an anterior inferior cerebellar artery loop under a nerve root near the pons.

A rare form of multiple exostosis with compression of the facial and trigeminal roots by prominent bone ridges is presented and the placing of a neuroprotector on trigeminal root is shown (Fig. 15).

The criteria for probable pathological neurovascular contacts are described in Table 4 and illustrated in Fig. 16.

Our clinical experience with neurovascular decompression is too small for any conclusions to be drawn. However, it seems that venous compression may be a causative factor. The veins fixing the trigeminal root at the bottom or behind assist pulsing arterial compression at the top or in front of the nerve (Fig. 17).

The main problems of endoscopic neurovascular disconnection are:

1. severance of vessels in conditions of coarse adhesions;

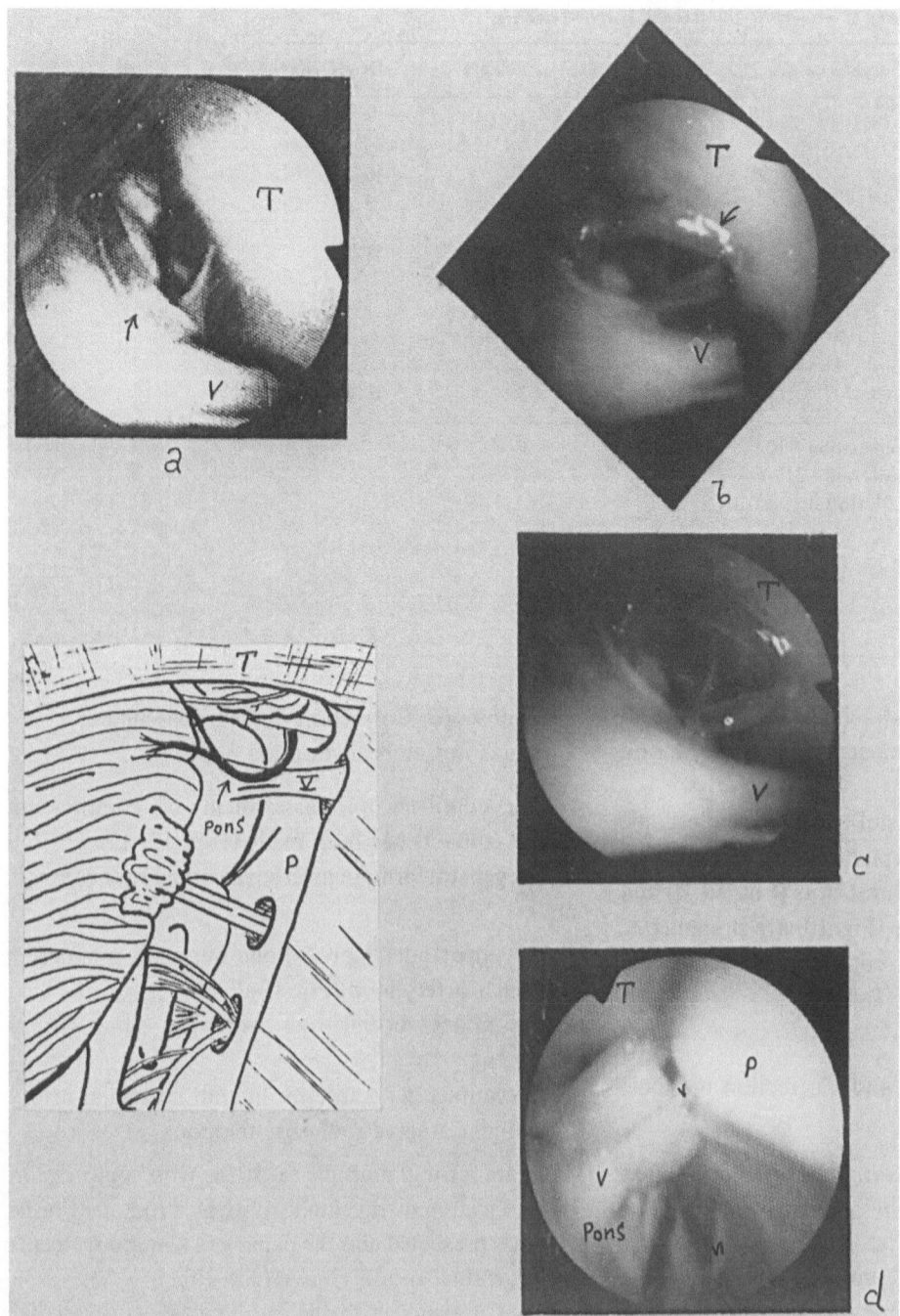

Fig. 12. Stages of endoscopic neurovascular severance (right cerebello-pontine angle). a) Contact discovery-upper lateral loop of ACS on the trigeminal root, b) loop mobilization by dissector, c) final nerve/vessel state, d) hollow microneuroprotector is placed on the nerve root

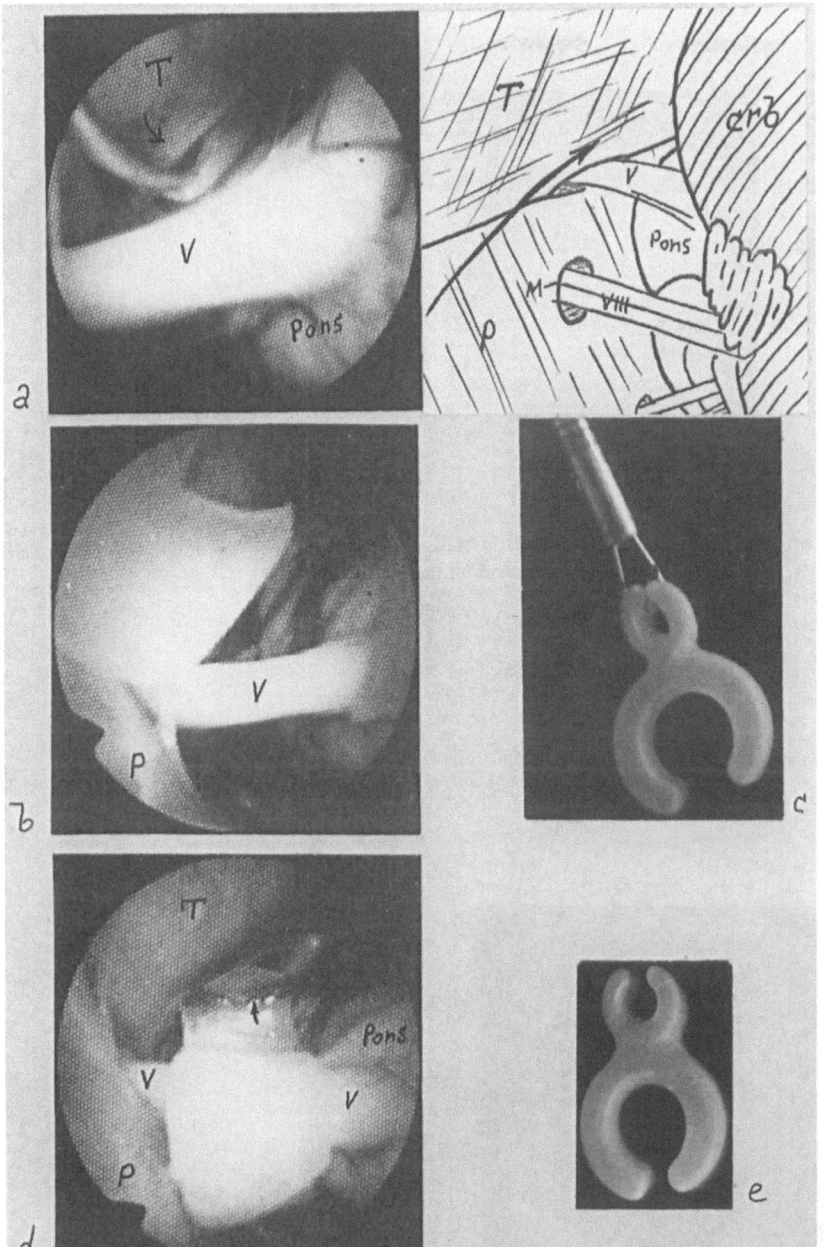

Fig. 13. The technique of neuroprotector placing on the left trigeminal root. a) Dissector severance of the upper medial ACS loop from nerve, b, c) long protector with open prongs, d, e) final topography and shape of neuroprotector; arterial loop is inserted between the short prongs

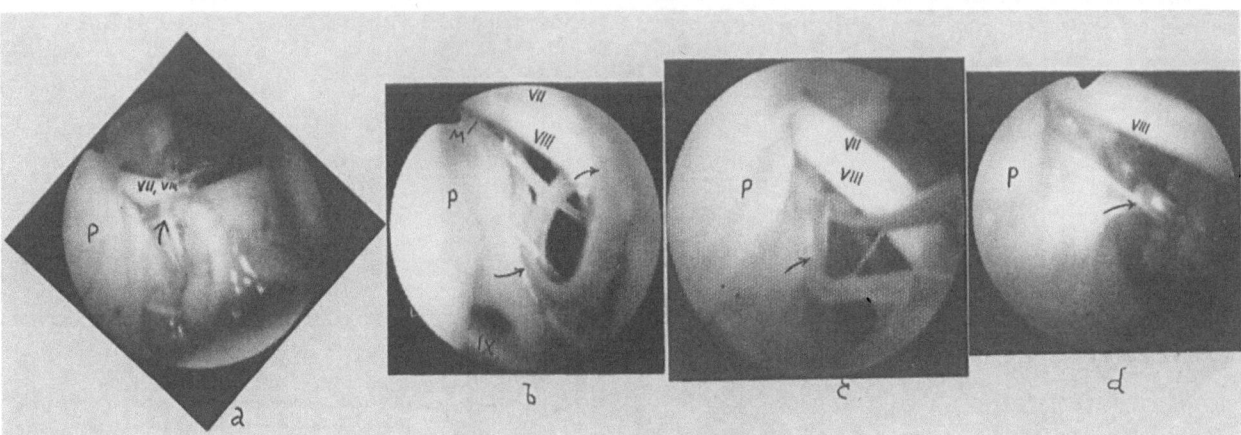

Fig. 14. Stages of endoscopic severance of ACIA loop from N.VII for placing of vascular neuroprotector. a) Contact discovery—upper lateral loop of ACS on the trigeminal root, b) loop mobilization by dissector, c) final nerve/vessel state, d) hollow microneuroprotector is placed on the nerve root

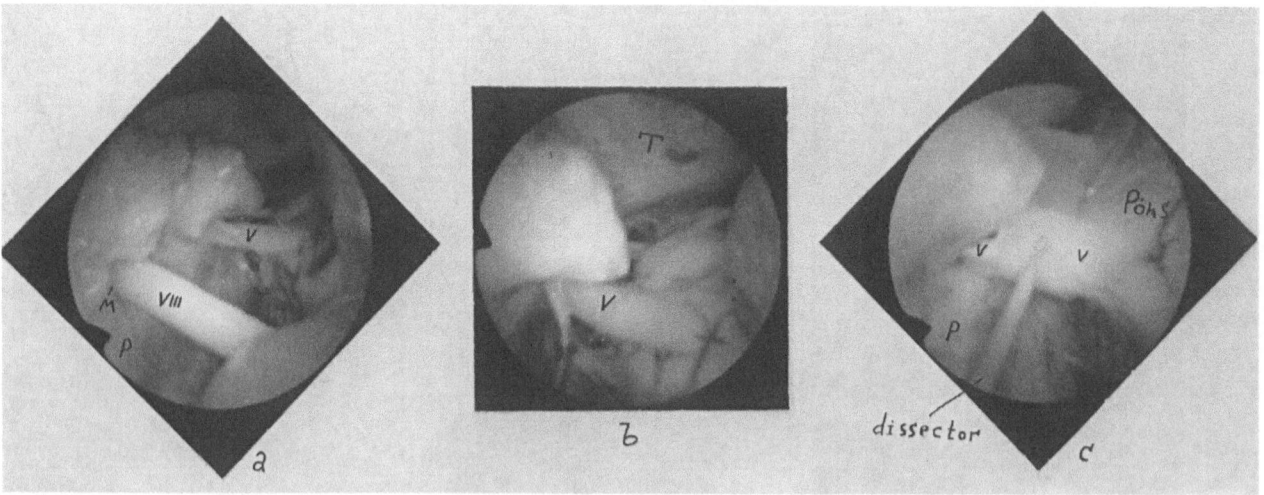

Fig. 15. Multiple osteoneural contacts in the left cerebellopontine angle. Posterior wall of the petrous temporal bone at its apex with exostoses compressing N. VII, N. VIII (a) and N. V (a, b). c) Neuroprotector placed on the N. V.

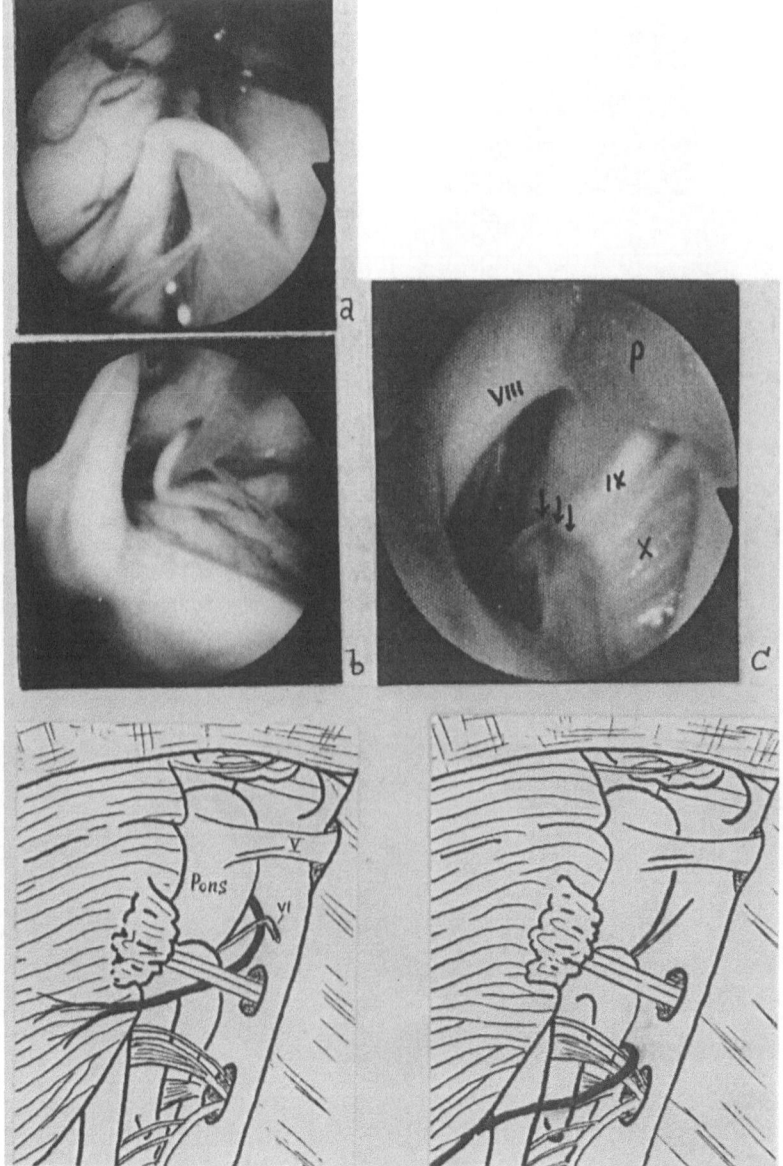

Fig. 16. Endoscopic criteria for neurovascular compression. a,b) A coarse axial deformation of the abducent nerve by ACIA loop (arrow), c) space disappearance between N.IX and N.X results from N.IX/ACIP loop contact

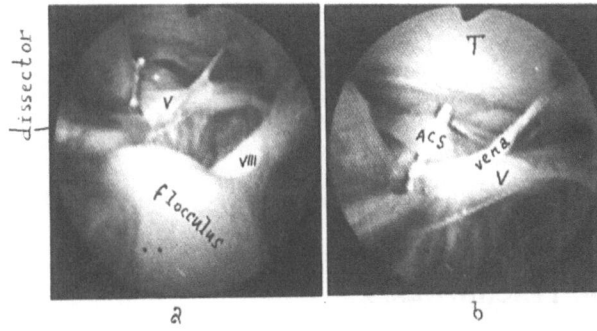

Fig. 17. Combined arterial (ACS) and venous compression of the right trigeminal root. Root fixation by the vein assisting arterial compression. Endoscopic dissector insertion (a) and upper medial ACS loop mobilization (b)

Table 4. *Endoscopic Criteria for Probable Pathological Neurovascular Contacts*

1. Disappearance of adjacent nerve spaces
2. Proportionality of nerve and vessel diameters
3. Nerve edge deformation
4. Axial nerve deformation

Contact rate on 260 cerebellopontine angle specimens (nonselected cadavers, mean age 67.4)

V/ACS – 27	VII/ACIA – 12
V/ACIA – 4	VIII/ACIA – 3
V/AB – 1	VII/ACIP, AV – 6
V/VPS – 16	IX/ACIA, ACIP – 4
VI/ACIA, AV – 5	

ACS – arteria cerebelli superior	AB – arteria basilaris.
ACIA – arteria cerebelli inferior anterior.	AV – arteria vertebralis.
ACIP – arteria cerebelli inferior posterior.	VPS – vena petrosa superior.

2. absence of bipolar coagulators for the coagulation of blocking petrosal veins in the approach.

Discussion

Recent advances of intracranial endoscopy as described in this paper, allow to investigate the precise natural stereotopography of supra- and infratentorial areas.

With the development and application of flexible endoscopes we succeeded in an extended visual investigation and description of the preformed intracranial cavities and the surrounding areas.

This endoscopic-topographic neuroanatomy represents the basis of new neurosurgical operative techniques under endoscopic control:

1. The evacuation of haematomas.
2. The assessment and correction of acute tentorial pressure cone.
3. The correction of neurovascular contacts in the cerebello-pontine angle.

At the moment, the transfer of the experimental results into clinical routine has to be considered critically.

First results of endoscopic evacuation of subdural and intracerebral haematomas are encouraging.

Much progress has been reached regarding the development of improved endoscopes. But special endoscopic instruments, especially suitable bipolar coagulators, are still missing. The construction of these instruments is demanded urgently.

The indications for endoscopic-neurosurgical interventions have to be defined in future. This has to be done with the evaluation of the experiences and results of other working-groups.

In conclusion endoscopy in neurosurgery gains in importance. The basis for application of endoscopic techniques is the endofiberscopic intracranial topographical anatomy.

References

1. Karakhan VB, Shuvaev KY (1988) Endoscopic anatomy of superior petrosal vein. Voprosy nejrokhirurgii 1:45–48
2. Karakhan VB, Shuvaev KY (1988) Endoscopic anatomy of acute tentorial pressure cone. Ibid. 5:50–55

Correspondence: V.B. Karakhan, M.D., Department of Neurology and Neurosurgery, Moscow Medical Stomatological Institute, 103473 Moscow, USSR. Department of Neurosurgery, Delegatskayu Str. 10/1.

Acta Neurochirurgica, Suppl. 54, 26–33 (1992)
© by Springer-Verlag 1992

Technological Fundamentals of Endoscopic Haemostasis

H.-D. Reidenbach

FH Köln/Cologne, Research Department for Biomedical Engineering/HLT, Köln, Federal Republic of Germany

Summary

In order to perform endoscopic haemostasis there exist several different mechanical, biochemical and thermal methods, which may be applied together with rigid or fully flexible endoscopes in different situations.

The technological fundamentals of convective, conductive and radiative heat transfer, the irradiation with coherent electromagnetic waves like microwaves and laser radiation and the resistive heating by RF-current are described.

A review of the state of the art of haemostatic coagulation by laser radiation (photocoagulation) and radio-frequency currents (surgical diathermy, high-frequency coagulation) is given.

The wavelength-dependent interactions of coherent light waves are compared especially for the three mainly different laser types, i.e. carbon-dioxide-, neodymium-YAG- and argon-ion-laser. The well-known disadvantages of the conventional RF-coagulation are overcome by the so-called electrohydrothermosation (EHT), i.e. the liquid-assisted application of resistive heating of biological tissues to perform haemostasis.

Different technological solutions for bipolar RF-coagulation probes including ball-tips and forceps are shown and the first experimental results are discussed in comparison.

Keywords: Endoscopic haemostasis; photocoagulation; laser radiation; high-frequency currents; electrohydrothermosation.

Introduction

Endoscopy provides methods and techniques for different medical fields today, which have been consequently developed further since the "invention for visualization of inner parts and diseases" by Bozzinti[3] from pure diagnostic use to operative and therapeutic applications.

It was in 1868 when Kussmaul was in urgent need of a sword-swallower to try his first rigid endoscope. The developments of semiflexible endoscopes by Schindler and Wolf obtainable since 1932 yielded to a much less dangerous diagnosis.

The first fully flexible endoscope was presented in 1957 by Hirschowitz, Peters and Curtiss, which initiated so many successful developments in gastroenterology.

Concerning for example the picture transmission systems there is a permanent increase in picture-quality when so-called coherent lightguides are used, but nevertheless things changed to an increasing use of high-tec semiconductor sensor chips. In this account an increasing number of different operative procedures may be guided by moving pictures on a screen rather than by a single mono- or stereoscopic view.

The state of the art in operative and therapeutic endoscopic techniques, especially in gastroenterology, gynaecology, arthroscopy and general surgery is given by several authors in a recently published book[4]. But there are further endoscopic applications such as in operative urology and neurosurgery for example, where a constantly increasing number of high-risk operations are performed with rigid and flexible endoscopes.

Most of the applications requiring complex surgical preparation need an efficent method to stop haemorrhages. Therefore haemostasis is at the centre of interest in endoscopic methods in the endocranial region, i.e. for trial biopsies in the brain and for extensive neurosurgical operations too. Sometimes there are analogies to "fishing in troubled waters".

Technological limits are set by the lumen of the working channel used together with the different biophysical principles in use.

Methods and Material

There are different possibilities for performing haemostasis in operative endoscopic procedures (Fig. 1). These can be divided into mechanical, chemical/biochemical and thermal methods[10]. For a comparison of the different methods of haemostasis Table 1 lists some criteria.

For endoscopic haemostasis thermal methods are especially

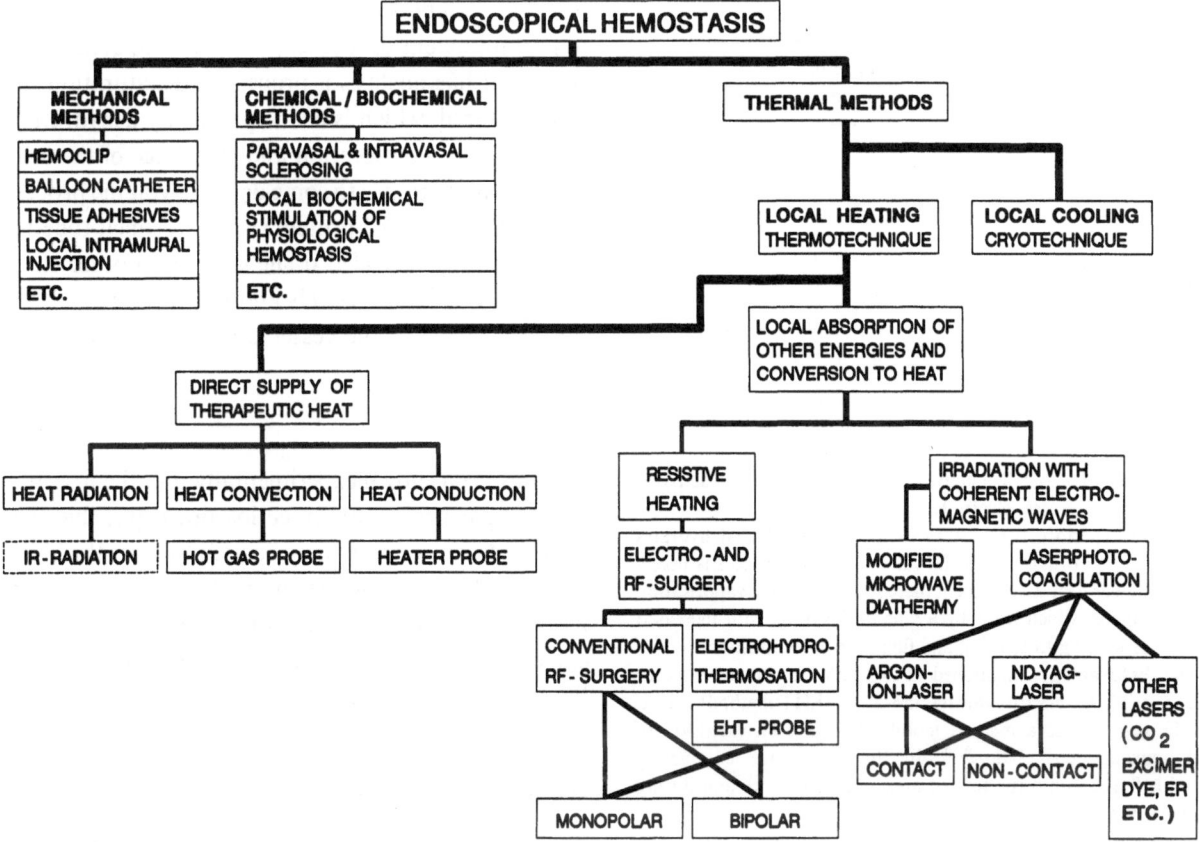

Fig. 1. Endoscopic methods for haemostasis

Table 1. *Criteria for Haemostasis*

- Primary rate of haemostasis
- Risk of recurrent haemorrhages
- Improvements in the surgery
- Practicability of the method
- Mobility
- Investment and maintenance
- Undesirable side-effects and risks

interesting, since there are many modalities to achieve an efficient and reproducible method of haemostasis. Quantitative investigations have been performed especially with high-frequency- and photo-coagulation[9].

Convective, Conductive and Radiative Heat Transfer

If a tiny tungsten helix is traversed by an electric current in a miniature probe and secondarily heats up a nitrogen jet to about 1500 °C, it is possible to arrest small bleeds by convection with such an inert gas stream directed onto the tissue surface[2].

Heating up a metallic probe by an electric current, the so-called "heater probe"[12] and the endocoagulator[13] can be realized.

Another possibility of exogenous heat delivery is given by heat radiation. In one such system a tungsten-halogen incandescent lamp with 250 W produces Infra-red-Radiation with a maximum wavelength of about 950 nm. The IR-Radiation is conducted via a solid lightguide onto the treatment site[7].

Haemostasis with Coherent Radiation

Non-ionizing, coherent electromagnetic waves include especially microwave- and laser-radiation.

Using the principle of light-amplification by stimulated emission of radiation which is realized in a LASER a monochromatic, nearly parallel (collimated) beam may be produced. This beam can be focussed by lenses to an extreme intensity, which is far greater than any other light-source available. The laser-radiation is emitted continuously or in a pulsed mode with high peak-powers during more or less short emission durations.

When endoscopic haemostasis is considered the application systems, i.e. the delivery system for laser radiation, is of great importance. Principally radiation may be transmitted by an articulated arm, consisting of optical components such as mirrors, lenses or prisms, which is movable in all three directions in space or by a fully flexible lightguide.

Beneath the so-called non-contact laser photocoagulation the contact method has been developed. In this situation different shaped probes made of sapphire or glass-ceramics are fixed to the distal end of the lightguide and brought into contact with the tissue to perform the desired haemostasis.

Haemostasis by Resistive Heating

A further possibility for performing endogenous heat generation is given when the temperature increase is used, which results from an electric current flowing through resistance measured in ohms. This principle of resistive heating is used in high-frequency surgery.

In the monopolar method a high-frequency current with a frequency between approximately 300 kHz and about 2 MHz, and in special cases up to approximately 10 MHz, "flows into" the tissue

Table 2. *Gradual Heat Reactions*

Kind of reaction	Temperature range or start of reaction (in °C)
● Hyperaemia ● Oedema	37–60
● Coagulation ● Boiling of tissue	Approx. 60
● Dehydration ● Necrosis ● Scurfing	Approx. 95
● Carbonization	Approx. 250
● Gaseous state	Several hundred

with the help of a so-called active electrode and "leaves" the tissue at a large-sized neutral electrode.

In order to perform haemostasis a temperature, which dehydrates the tissue in accordance with the values given in Table 2 is necessary. At higher heat development there is a progressive vaporisation of tissue fluids, which leads by a glue-like process with increased adherence tending to scurfing and finally to carbonization.

The disadvantages mentioned above may be overcome as far as is possible by electrohydrothermosation (EHT-method), which can be regarded as a method modification and an extension of high-frequency coagulation[9,11]. In this case poor conducting liquid is flushed onto the treated area of tissue before and during the current-flow.

The so-called salt-jet-method may be regarded as an option of the EHT-technique[9]. In this case for example a physiological, that is an isotonic sodium-saline-solution is flushed onto the surface of the tissue to be treated at a flow-rate of about 0.4 to 1 ml/s. The high-frequency current is directed to the tissue via the salt-jet, and thus the high-frequency coagulation is in some respect contactless.

Results

General Heat Effects

Special interest belongs to the methods of locally coagulative heat production. With this the heat-dependent coagulation of proteins, i.e. the transition of colloidal material from the brine-state into the gel-state, is a matter of fact for haemostasis. Denatured proteins are the result of such an irreversible process. The total event of haemostasis is a much more complicated process than may be described by mere denaturation, since especially occlusion of a vessel should happen. The heat, which is produced exogenously and endogenously, produces vascular wall contraction, an aggregation of erythrocytes and thrombocytes, and the activation of fibrinizing enzymes to an intravascular blood-clot resulting in physiological haemostasis.

Principally different reactions may be arrived at by increasing degrees of heat (Table 2).

The listed reactions to varying degrees of heat and temperature ranges or starting points are dependent to some extent on the different tissues but to an important extent on the length of time of action. For successful haemostasis a process of denaturation has to be induced, which is accompanied by dehydration. This leads to a desiccation and shrinkage of the cells.

The coagulation is to some degree a volume process, where the denatured protein substances melt together thus delivering a sealed tissue surface. In other cases, a contraction of the vessels and the development of a thrombus fixed to the vessel is induced by heat.

Hot Gas Probe

It has been shown experimentally that the insertion of a hot gas probe into an endoscope is possible. The flushing hot gas stream may be used to blow the blood away in this non-contact method, but the thermally induced coagulum is too porous as a result of an extensive carbonization, and therefore produces only limited haemostatic effectiveness.

Heater-probe and Endocoagulation

The heat-contact probe has to be pressed firm onto the tissue, since the heat transfer happens by conduction. The adherence of coagulated biological tissue can be reduced through a fluor-polymer-cladding of the contact-tip. A further reduction may be accomplished by simultaneous liquid-irrigation.

In the case of an endocoagulator the stickiness is reduced by limiting the temperature of the probe to approx. 110 °C. Both probes are applied endoscopically, especially for gastrointestinal haemorrhages and for haemostasis of small vessels in the abdomen.

IR-Coagulator

Haemostasis is achieved by absorption of the radiation, i.e. by photocoagulation, when the distal tip of the lightguide is pressed onto the tissue simultaneously. Adhesion is reduced by using antiadhesive caps made from PTFE or PFA or by the use of highly heat-conducting and high-melting sapphire crystals fixed to the lightguide-tip.

Laser

In order to perform photocoagulation the radiant energy from the optical spectrum is converted into heat by the process of absorption. The absorption of photons themselves is a more or less wavelength spe-

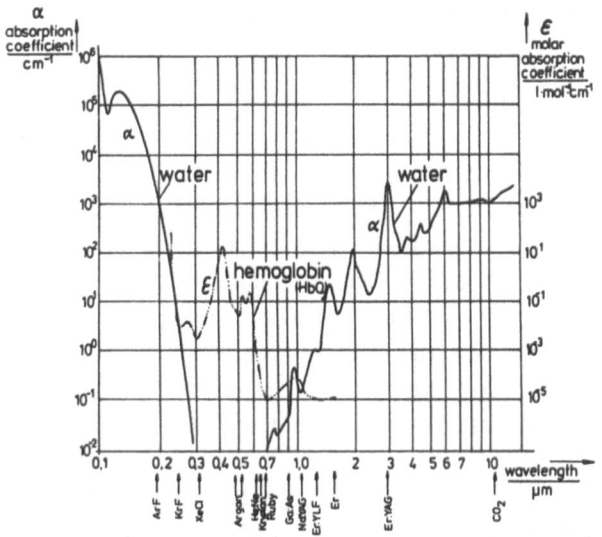

Fig. 2. Absorption of haemoglobin and water as a function of wavelength (after Boulnois, J.-L.: Lasers in Med. Sc. *1* (1986), 47)

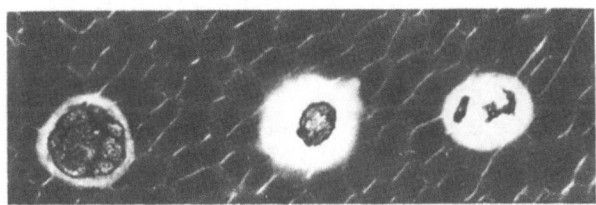

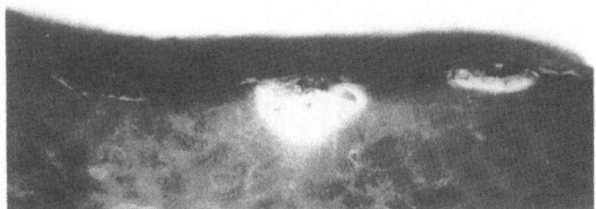

Fig. 3. Laser coagulations on freshly butchered liver tissue from pigs; *above*: frontal view; *below*: incision; *from left to right*: CO_2-laser (approx. 25 W), Nd:YAG-laser (approx. 50 W), argon-ion-laser (approx. 5 W)

cific and selective process, which occurs at an atomic or molecular level and is influenced especially by the absorption of haemoglobin and water (Fig. 2).

Apart from absorption the interaction of laser light is determined by the scattering behaviour in biological tissues. Figure 3 shows an example of the various rates of absorption and scattering from liver-tissue at 3 typical laser wavelengths, i.e. at $10.6\,\mu m$ (CO_2-laser), $1.064\,\mu m$ (Nd:YAG-laser), and approx. $0.5\,\mu m$ (argon-ion-laser).

The degree and extension of coagulation may be varied by the physical parameters of power, exposure time, irradiated area, and kind of operation, i.e. continuous or pulsed mode at the before mentioned wavelengths[11].

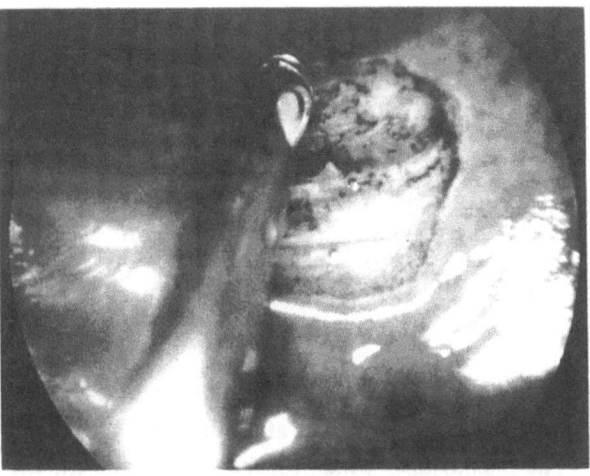

Fig. 4. Endoscopic surgery of the bowel mucosa with CO_2-laser radiation and the use of a 90°-beam deflector

Since at this time the developments of flexible fibers with a sufficient power durability at $10.6\,\mu m$ are in their infancy – probably AgCl-crystals or something else may be a promising material for the production of infra-red fibers – CO_2-laser radiation is mainly directed via a microscope adaptor, (which should be interesting in neurosurgery), or is guided through articulated arms to the site of operation.

In addition hollow waveguides made from ceramics or glass-ceramics with or without a ninety-degree beam deflector can be used for surgery, where the excellent cutting capabilities of CO_2-laser radiation can be exploited (Fig. 4).

As far as argon-ion- and Nd:YAG-lasers are concerned lightguided systems are available with adequate properties. Besides the chosen lightguide a coaxial inert gas insufflation is of great importance for a laser photocoagulation system. By this the bleeding site can be blown free from overlying blood and the efficiency of the laser interaction is greatly increased. Moreover an unneccessary carbonization of the tissue and a destruction of the distal end of the lightguide resulting from splashed up fragments which are highly absorbing, is prevented. The common distance between lightguide and tissue surface for performing haemostatic coagulation ranges from 1 to 2 cm. A greater distance does not result in haemostasis, while at a shorter distance vaporisation of tissue occurs, for example when a Nd:YAG-laser is used.

Conventional RF(High-Frequency)-Coagulation

The greatest current density is under the small-sized active electrode, where the electric resistance is at

Table 3. *Disadvantages of the Conventional High-Frequency Coagulation Method*

- No visibility of the real source of haemorrhage
- Adherence of the coagulum to the probe
- Provocation of recurrent haemorrhage when the probe is removed
- Coagulation depth is difficult to control in bleeding situations

maximum in the closed circuit, thus resulting in a maximal heat development at this point[9, pp. 126ff].

The haemostasis which can be arrived at with the conventional method using a high-frequency electrode suffers from the following disadvantages (Table 3), which are partly shown in Fig. 5. A typical event happens, when high-frequency coagulation has to be performed in a nearly blind fashion.

Electrohydrothermosation (EHT)

By this liquid instillation the source of a haemorrhage can be localizied initially (Fig. 6). Coagulation follows under simultaneous liquid perfusion. That means firstly that issuing blood may be flushed away and secondly that the temperature of the surface is limited to approximately 100 °C owing to the cooling effect of the instilled fluid. Simultaneously an undesired transfer of heat from the tissue to the electrode is avoided, since the temperature-range, where a glutinous-adhesive state supervenes, is not reached at all. The advantages of the EHT-method are listed in Table 4.

The manufacturing of bipolar EHT-probes for endoscopic purposes makes great demands owing to the smallness of the probes themselves. Figure 7 shows a vessel-occlusion, which has been produced with a prototype of an endoscopic bipolar probe.

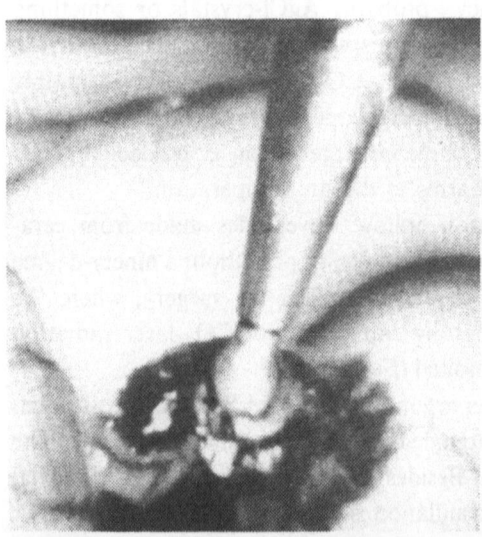
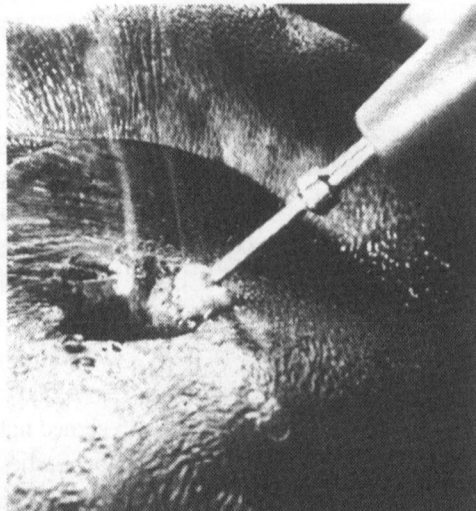

Fig. 5. Examples of disadvantages of the conventional high-frequency coagulations; *left*: no visibility due to foaming up blood; *right*: development of smoke and adherence of the probe

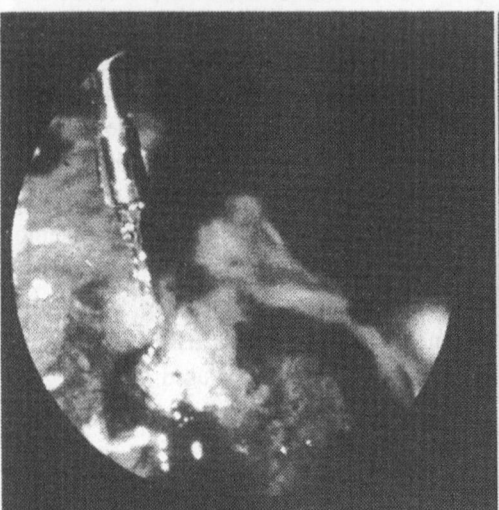
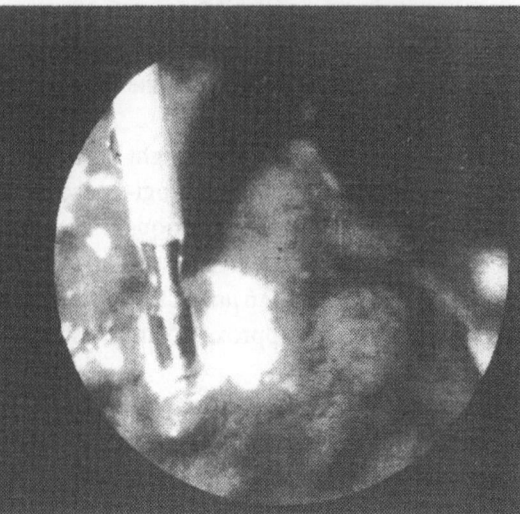

Fig. 6. EHT-method; *left*: rinsing of the bleeding tissue; *right*: coagulation under simultaneous liquid perfusion

Table 4. *Advantages of EHT-Coagulation*

- Nearly bloodfree tissue surface
- Improved visibility
- No adhesions to the probe
- Calculable depth of coagulation
- Reduced duration of treatment

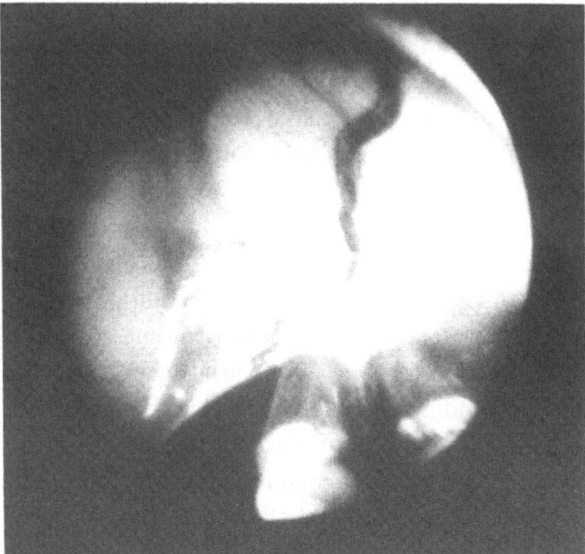

Fig. 8. Coagulation of a vessel with bipolar EHT-forceps

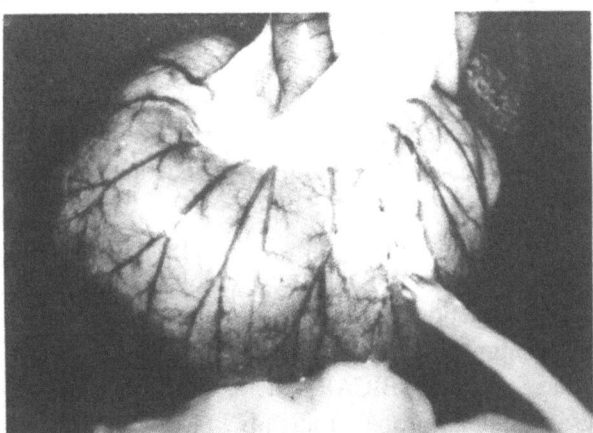

Fig. 7. Vessel occlusion produced with a bipolar EHT-probe

Besides this cylindrically-shaped bipolar probe bipolar forceps have been produced at an experimental stage. They have already been used to coagulate vessels of the oesophagus at endoscopic microsurgical dissection of this organ (Fig. 8).

Discussion

The methods applied are dependent on the particular indications since haemorrhages belong to different conditions. When a spontaneous haemorrhage occurs during an endoscopically performed operation there is a need for a very quick improvement in visibility. On the other hand, if we are talking about microsurgical operations, as for example with minimal invasive surgery, the haemostatic procedure has to be adequate during the operation, but need not be too definitive until the end of the operation.

There is an enormous range of surgical bleeding, from capillary to dramatically spurting arterial haemorrhages, with or without a visible vessel stump. Therefore different ways to fulfill all these demands must be considered especially with respect to the saving of time.

In the special case of a very grave haemorrhage an optimum in primary haemostasis rate with a simultaneous minimal risk of residual bleeding must be achieved. Last but not least the performance/price-

relation will be decisive when the effectiveness of various methods is compared.

The coagulum achieved by hot gas probes is strongly carbonized and too porous to produce a sufficient haemostatic result[2]. With heater-probes and endo-coagulator-tips sufficient coagulation may be obtained, when the haemorrhage is not too severe.[12,13]. Endoscopic use of incoherent IR-Radiation, which penetrates comparatively deep into the tissue, is physically impossible in connection with a working channel of about 2 mm in a fully flexible endoscope. It is not possible to collect the necessary power density with incoherent infrared-radiation and other artificial radiation sources.

With microwaves at the ISM-frequency of 2.45 GHz the potential of haemostasis has been demonstrated[1]. The problems of power handling capabilities with flexible waveguides and the geometrical dimensions of the antennae needed for achieving haemostasis prohibit endoscopic application at this frequency[9]. At mm-waves the geometrical limitations are reduced, but there are no powerful mm-wave-generators and waveguides available now.

As a result of the relatively high absorption coefficient for laser radiation at 10.6 μm in water, the tissue-absorption is very high. There is only a superficial reaction. This produces a necrotic scurf even at small power densities and short times of exposure. On the other hand, the coagulation effect is almost negligible, which means that CO_2-laser radiation is useful only for haemostasis of small vessels or superficial oozing.

Resulting from the approximately 4 to 5 times

higher penetration of radiation from a Nd:YAG-laser at $1.064\,\mu m$ in comparision to the argon-ion-laser radiation (cf. Fig. 2) in tissue with a high blood content, there is a more pronounced volume absorption at higher radiant flux. This is supported by the relatively strong scattering (Fig. 3) at this wavelength in biological tissues.

The scattering effect is increased even more when tissue whitens as a result of protein denaturation. Thus the effect of coagulative haemostasis using Nd:YAG-laser radiation is the result of shrinkage of vessels in the surrounding tissue due to successive occlusion. In parallel there is haemostasis by direct thrombosis.

The radiation of an argon-ion-laser in the green-blue spectrum is nearly selectively absorbed in blood (Fig. 2), while the absorption in water is comparably small. As a result there is superficial volume absorption at once, which produces carbonization especially at longer exposure times. The photocoagulation-efficacy of the argon-ion-laser had already been shown in 1975, when the first successful endoscopic haemostasis in the gastrointestinal tract was reported[5].

Photocoagulation may also be performed with so-called contact-tips. The laser power necessary in this case may obviously be reduced.

For a comparison of the both methods one has to consider that the up to now promised advantage of a non-contact method is lost when the contact method is applied for haemostasis. Laser photocoagulation has shown its efficacy in situations where haemostasis is the aim of treatment.

The well-known RF-diathermy or high-frequency coagulation method is another modality for achieving haemostasis.

The coagulation depth which may be produced on a dry tissue surface is nearly equal to the diameter of the applied electrode. Since the heat generation is the result of a current which flows into the tissue, the coagulated volume shows some inhomogeneity, especially when there are vessels lying in the interaction zone, because vessels show a greater electrolytic conductivity. Therefore the accompanying temperature-field shows a more or less frayed coagulation border-line, which differs from the relative smooth border in laser- or infra-red-photocoagulation.

The disadvantages of conventional high-frequency coagulation are well known. Thus in practice, mechanical and electrical contact of the active electrode to the forceps or the suction instrument, are necessary . The bleeding vessel is grasped by forceps and the active electrode is brought into close contact with one prong of the applied forceps. In this way coagulation by adaptation results. But it is possible today to arrange for suction and electrohydrothermosation to be combined in one instrument.

Distilled water has been used as a non-isotonic flushing fluid and no side-effects have so far been reported. To prevent any disadvantages during neurosurgery as a consequence of hypotonia, isotonic conditions may be provided by simply adding of approximately 5% glycerine. Alternatively a 5% glucose-solution may be used, but problems eventually arise due to sticking together of the nozzles, which are necessary for the delivery of the liquid, and finally of the total delivery system.

The bipolar method has to be applied – whenever possible – to prevent aberrant currents causing undesired reactions away from the area of treatment. For example in gastro-intestinal haemostasis a multipolar, so-called BICAP-system, an endoscopically usable bipolar probe[6], is already in use. Experimental results have shown, that there are some problems concerning the availability, as a result of thermal and corrosive side effects.

In the case of "salt-jet-probes" haemostasis has been achieved experimentally. The amount of fluid, which is necessary, and the resulting depot-like "electrolytic lake" prohibit in unfavourable cases an efficient application of the salt-jet, for only the unimpeded salt-jet will achieve effective coagulation.

When exogenous and endogenous heat is used different physical modalities for the induction of coagulation with subsequent haemostasis are called into play. Laser- and high-frequency coagulation are especially in use in various medical fields. By varying the mode of operation (continuous or pulsed wave), power, power density (the distance between lightguide and tissue and the geometrical dimensions of the probe), duration of use and not at least simultaneous support, i.e. gas insufflation or fluid irrigation, different reactions to haemostasis are arrived at. A final comparison of the efficacy in clinical neurosurgery must include the costs of the particular method.

References

1. Bodem F, Reidenbach H-D, Nehls M, Brand H, Frühmorgen P (1975) Untersuchungen über die Eignung der 2,45 GHz-Mikrowellenerwärmung für eine endoskopische Koagulationstherapie. Biomed Techn 20: 238–239
2. Bodem F, Reidenbach H-D, Frühmorgen P, Matek W, Kaduk B (1980) Investigation on the haemostatic efficacy of the thermocoagulation of gastrointestinal hemorrhages by convective

heat transfer via a miniature endoscopic hot gas probe. Biomed Techn 25: 179–181

3. Bozzini P (1806) Lichtleiter, eine Erfindung zur Anschauung innerer Teile und Krankheiten. J prakt Heilkd 24: 107–125

4. Buess G (Hrsg) (1990) Endoskopie. Deutscher Ärzteverlag, Köln

5. Frühmorgen P, Demling L, Bodem F, Reidenbach H-D, Brand H, Kaduk B (1975) Erste endoskopische Laser-Koagulationen im Gastrointestinaltrakt des Menschen. Dtsch med Wochenschr 100: 1678

6. Laime L (1987) Multipolar electrocoagulation in the treatment of acute upper gastrointestinal tract hemorrhage. N Engl J Med 316: 1613

7. Nath G, Kreitmair A, Kießhaber P, Moritz L (1978) Neue Infrarot-Koagulationsmethode. In: Rösch W (Hrsg) Fortschritte in der Endoskopie. Perimed, Erlangen, S 17–19

8. Reidenbach H-D, Bodem F, Frühmorgen P, Schroeder G, Lex P, Kaduk B (1978) Eine neue Methode zur endoskopischen Hochfrequenzkoagulation von Schleimhautdefekten. Biomed Techn 23: 71–74

9. Reidenbach H-D (1983) Hochfrequenz- und Lasertechnik in der Medizin. Springer, Berlin Heidelberg New York

10. Reidenbach H-D (1990) Technologische Grundlagen der endoskopischen Blutstillung mit flexiblen Instrumenten. In: Buess G (Hrsg) Endoskopie. Deutscher Arzteverlag, Köln, S 105–124

11. Reidenbach H-D (1990) Grundlagen der Präparations- und Blutstillungstechnik. In: Buess G (Hrsg) Endoskopie. Deutscher Arzteverlag, Köln, S 231–243

12. Rubin C, Auth D, Silverstein F, Protell R, Dennis M (1976) Preliminary study of a teflon-coated heater for endoscopic control of upper gastrointestinal bleeding in a standard ulcer model. Gastrointest Endosc 22: 235

13. Semm K (1975) Elektronisch gesteuerter Schwachstrom als Ersatz des Hochfrequenzstromes. In: Ottenjann R (Hrsg) Fortschritte der Endoskopie, Bd. 6. Schattauer, Stuttgart, S 17–21

Correspondence: Prof. Dr.-Ing. H.-D. Reidenbach, Research department for biomedical engineering/H L T, FH Köln, Betzdorfer Strasse 2, D-W-5000 Köln 21, Federal Republic of Germany.

Acta Neurochirurgica, Suppl. 54, 34–41 (1992)
© by Springer-Verlag 1992

Ultrasound Stereotaxic Endoscopy in Neurosurgery

L. M. Auer

Neurosurgical Clinic, Saarland University Medical School, Homburg/Saar, Federal Republic of Germany

Summary

Stereotaxic endoscopy assisted by laser- and video technique can be used in a circumscribed number of intracerebral lesions. In the present series of patients, ultrasound imaging has been used as a stereotaxic method to guide the tip of the endoscope to a target area in the depth of the brain via a burr hole in the cranial vault. Initial experience has been obtained in the evacuation of various intracerebral and ventricular as well as cerebellar haematomas; moreover, ventricular tumours can be laser-coagulated and resected. In cases of cystic hemispheric tumours biopsies can be taken under visual control and the inner surface coagulated with laser.

The method is less traumatic compared to conventional neurosurgery.

Keywords: Ultrasound guided stereotaxy; neuroendoscopy; cerebral haematomas; ventricular tumours; cystic brain tumours.

Introduction

Both stereotaxy and endoscopy are nearly as old as neurosurgery itself. While stereotaxic techniques made a stepwise development with the introduction of new instruments and modern imaging techniques the indications for endoscopy remained limited owing to the impossibility of being able to guarantee sterile conditions during operation and to prevent bleeding in the confined operating field in the depth of the brain. In recent years, however, several technical developments have made the application of endoscopic techniques in neurosurgery again attractive: The prevention of bleeding is easier, with aid of laser energy. Moreover, the development of modern imaging techniques enabled stereotaxy and endoscopy to be combined for precise guiding of the endoscope from a burr hole to a target area in the depth of the brain. Thus, CT guided stereotaxy can be used for endoscopic procedures. Another imaging technique recently introduced into neurosurgery allows even intra-operative imaging of the brain either via a burr

hole or craniotomy, namely real-time echotomography (ultrasound imaging). Direct visual control of the operating field itself through the endoscope is then achieved by aid of video techniques.

Ultrasound Guided Stereotaxy

Real time echotomography or ultrasound imaging has been used in neurosurgery for about one decade. The ultrasound probe can be placed onto the brain surface via a craniotomy, which allows direct visual control of a surgical instrument introduced into the brain aside the ultrasound probe (simultaneous ultrasound stereotaxy) (Fig. 1 a). Alternatively, the ultrasound probe can be applied via a burr hole for imaging of the target area in the depth and selection of a target point as well as for calculation of the depth of penetration from brain surface to target point; while the direction to target is kept by a fixation device, the ultrasound probe is removed and replaced by a surgical instrument such as a biopay needle or an endoscope (consecutive ultrasound stereotaxy) (Fig. 1 b–d). Figure 2 shows an ultrasound probe adapted for consecutive ultrasound stereotaxy with the aid of a fixation ring device. Figure 3 shows an example of a ventricular tumour on MRI scan and on ultrasound images for comparison of spatial resolutions of these techniques. Many anatomical details can be seen on the ultrasound images.

State of the Art of the Neuro-Endoscope

The last step in the development is shown on Fig. 4: The Neuro-Endoscope[R] (Karl Storz Co. Tuttlingen) has an outer diameter of 6 mm. The rigid outer tube has a blunt oblique tip with a side hole. The purpose

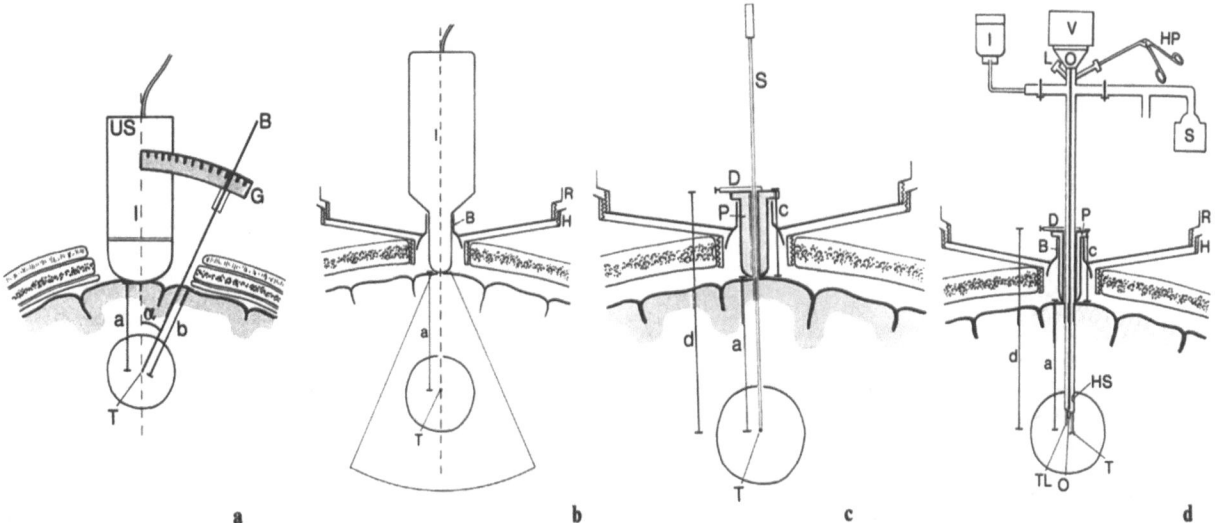

Fig. 1. a) Simultaneous ultrasound guided stereotaxy: *a* distance between brain surface and target point in the depth of the brain, α = angle between central ray and *B*, *b* distance between brain surface and target point along *B*, *B* biopsy needle, *G* instrument guide with indication of angle for angle α, *1* central ray of the ultrasound image, *T* target point, *US* ultrasound probe. b) Consecutive ultrasound stereotaxy: *a* distance between brain surface and target point, *B* ball joint, *H* fixation ring for the ultrasound probe and biopsy instrument, *1* central ray of ultrasound probe, *R* fixation ring, *T* target point. c) Consecutive ultrasound stereotaxy for biopsy via a burr hole: *a* distance between brain surface and target point, *c* length of guide *P*, *d* distance between needle guide and target point (a + c), *P* guide for biopsy needle, *S* biopsy needle, *T* target point. d) Ultrasound stereotactic endoscopic microneurosurgery: *H* and *R* fixation ring for ultrasound probe or neuro-endoscope. *I* infusion system for the neuroendoscope, *L* Neodym YAG laser connector, *O* and *V* connector for video system, *S* suction system, *D* tip of microbiopsy forceps, *HP* microbiopsy forceps, *TL* tip of microlaser tube

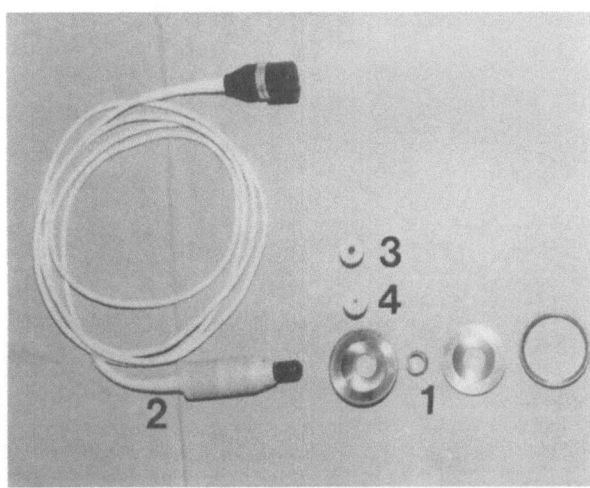

Fig. 2. Ultrasound set for consecutive ultrasound stereotaxy: *1* fixation device, ball joint and fixation ring, *2* ultrasound head, *3* guide for endoscope, *4* guide for biopsy

of this design is to obtain a clear view of the operating field. Inside, the endoscope is separated into two half-tubes, which have channels for the optic system, the irrigation and suction system, a micro-instrument such as forceps or scissors; in addition, one channel harbours an optic fiber for the application of Neodym-Yag laser energy. The connector for the laser fiber harbours a spring device which allows some forward-backward movement of this fiber in the

operative field. The Hopkin's optics are available for straight forward view, but also angled between 30 and 135° to allow a view to the side or even backwards. A video camera is connected to the ocular lens. The surgeon operates by the television screen; this circumstance warrants a sterile operative field, because the surgeon does not need to look through the ocular lens. One semi-tube is used to bring warm Ringer solution or artificial CSF over the optic lens into the surgical cavity. The oblique shape of the tip of the endoscope and the side-hole on the suction channel guides the irrigation outflow stained with blood or detritus as far away from the optic lens as possible. One must consider, that the whole operation is performed under water, therefore it is of crucial importance to have clear irrigation fluid in the immediate surroundings of the endoscope's tip in the operating field. The Neodymium yttrium aluminium garment (Neodym-Yag) laser (Messerschmidt-Bölkow-Blohm, München, Germany) is used at an energy level of 20–40 W; it serves for coagulation of small surface vessels of a tumour or the wall of a haematoma cavity.

Intracranial pressure during operation can be regulated by adjusting the outflow tube of the irrigation system.

In case further imaging control by aid of ultra-

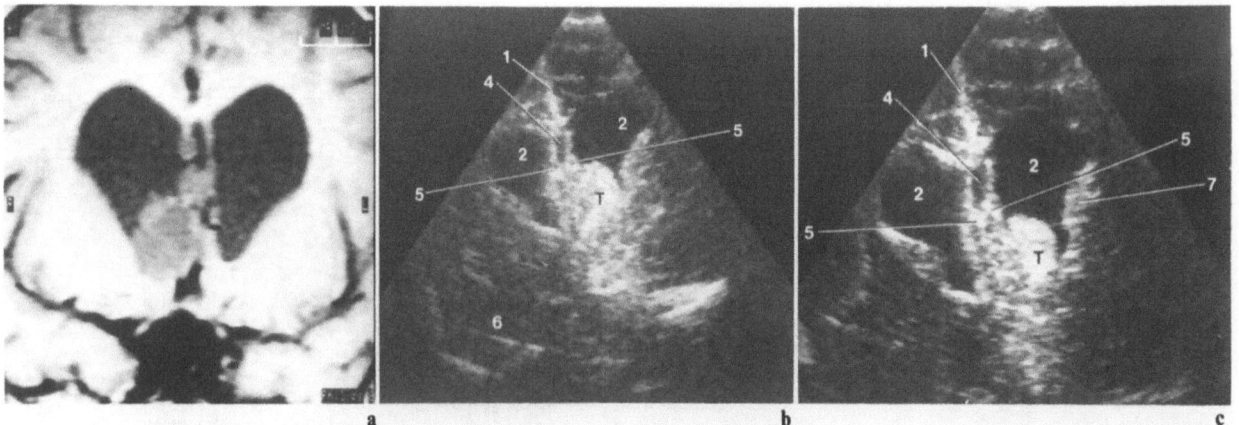

Fig. 3. Intraventricular low grade astrocytoma. a) Preoperative MRI, coronal section. b) and c) Intraoperative ultrasound image, coronal section with 5 MHz ultrasound head: *1* interhemispheric fissure, *2* lateral ventricle, *3* third ventricle, *4* cavum vergae, *5* columnae fornicis, *6* left temporal fossa, *7* caudate nucleus. *T* tumour

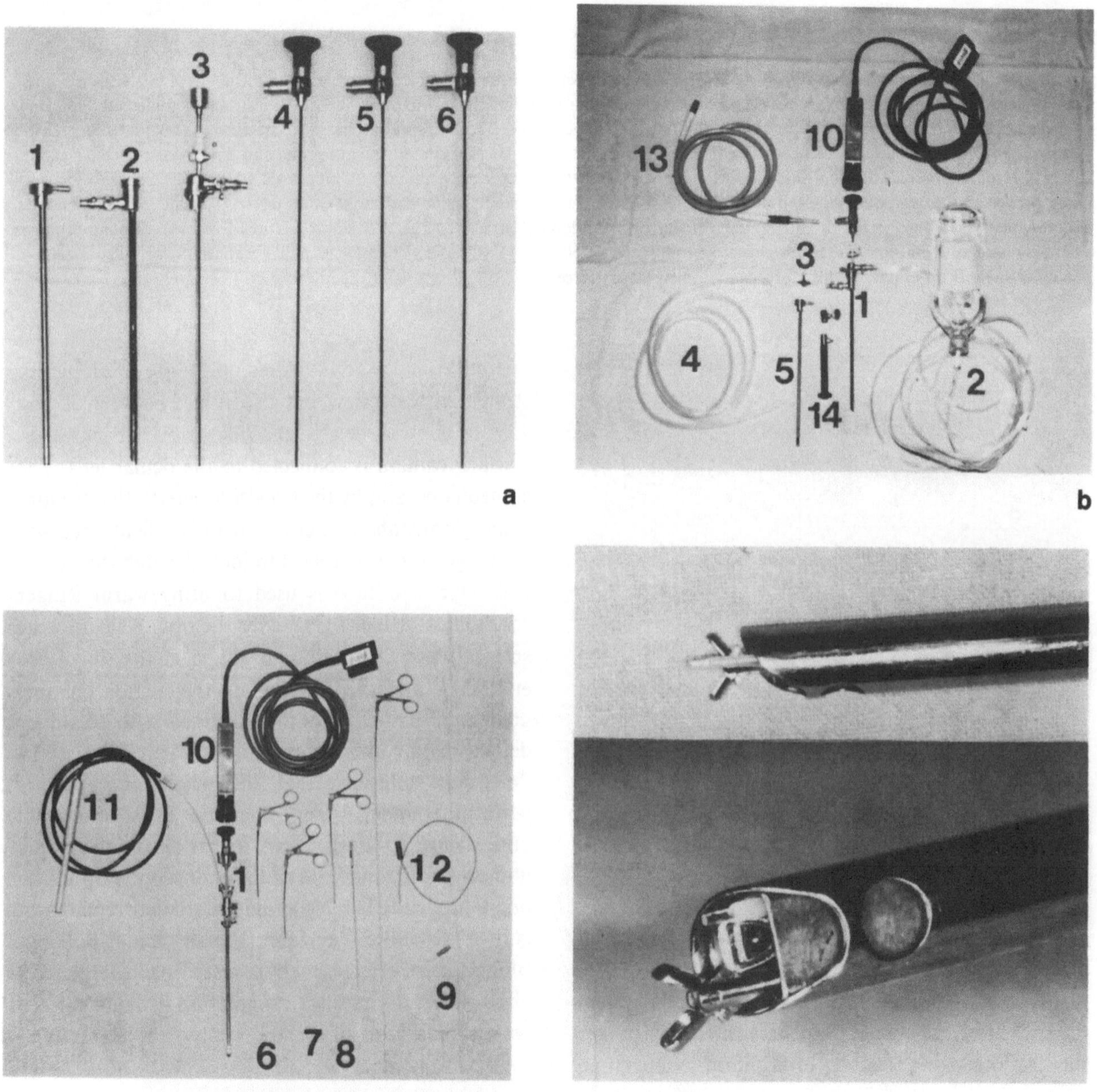

Fig. 4. Instruments for neuro-endoscopy: a) *1–6* parts of the endoscope, b) and c) *1* neuro-endoscope, *2* irrigation system for endoscope, *3* and *4* suction system, *5* mandrin, *6* biopsy forceps, *7* micro-scissors, *8* forceps, *9* suction tube, *10* video camera, *11* Neodym YAG laser tube, *12* monopolar coagulation probe, *13* cable for cold light source, *14* safety handle, d) tip of endoscope with oblique end and additional suction channel, optic system, biopsy forceps and tip of laser probe

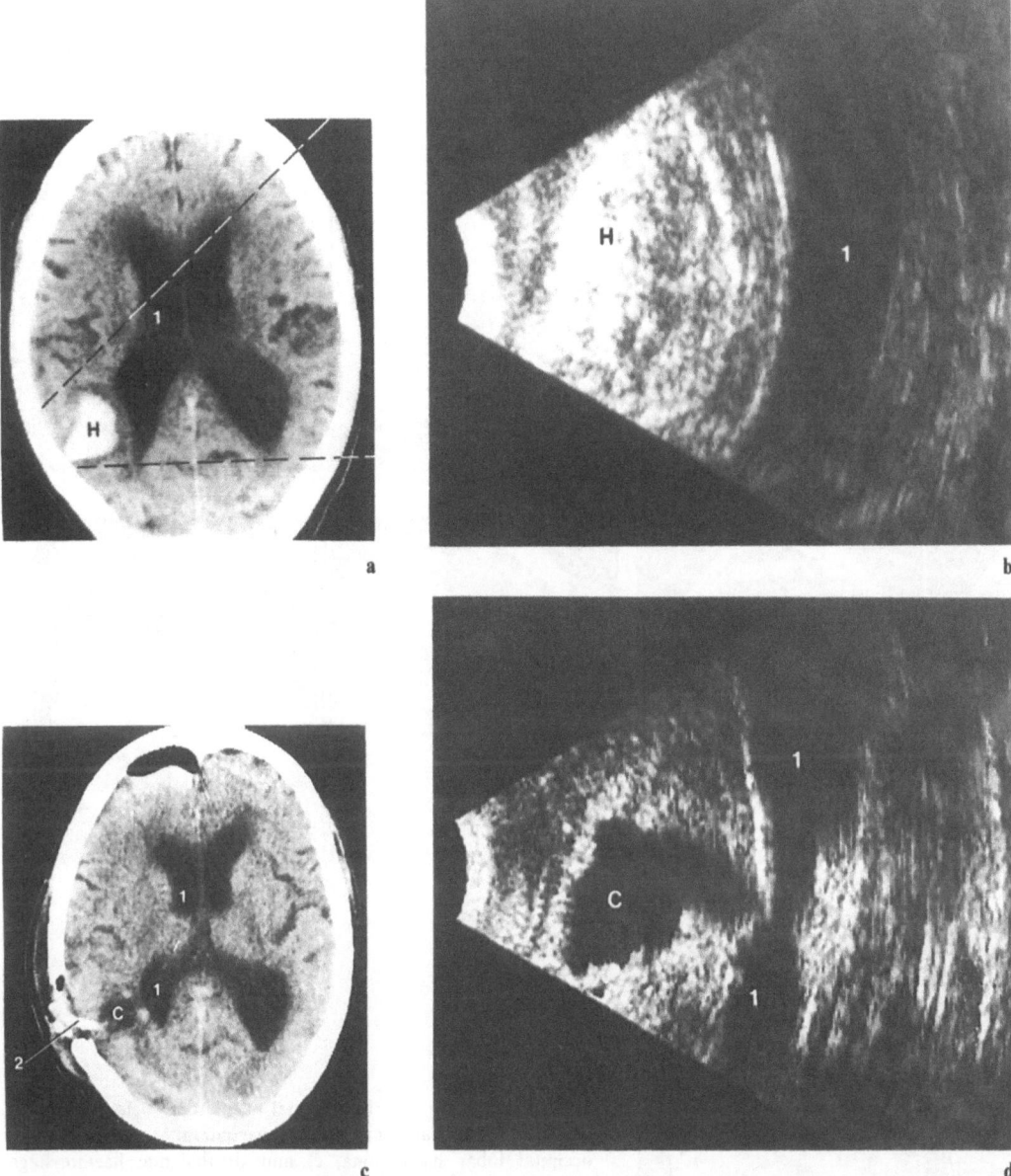

Fig. 5. Left temporal occipital haematoma in a 72 year old male patient: a) preoperative computerized tomogram, b) intraoperative ultrasound image with 7.5 MHz-probe before evacuation of the haematoma, c) intraoperative ultrasound image following evacuation of the haematoma, d) postoperative control computerized tomogram; *1* left lateral ventricle, *2* drainage catheter in the empty haematoma cavity, *H* haematoma, *C* empty haematoma cavity

sound is wanted or needed during operation, the endo-scope is removed from the stereotaxic holder, and the ultrasound probe is re-introduced.

The method presented here has been used in three different types of cerebral lesions: intracranial haematomas, ventricular tumours, cystic tumours of the cerebral hemispheres.

Cerebral Haematomas

Surgical treatment of intracerebral haematomas is under debate; however, it may be considered in some cases. A conventional method for evacuation of an intracerebral haematoma consists of a craniotomy and exposure of the haematoma via corticotomy. Endoscopic evacuation can be done via a burr hole. Liquid blood can easily be washed out with the aid of the irrigation suction system, coagulated blood is evacuated either by using a mild suction system at the outflow side of the endoscope or by additional morcelation of clots with the aid of biopsy forceps. The goal of the endoscopic procedure is reduction of intracranial pressure produced by the mass without emphasis on radical evacuation of clots. Above all touching the wall of the haematoma cavity with instruments must be avoided; it is only the fluid flow

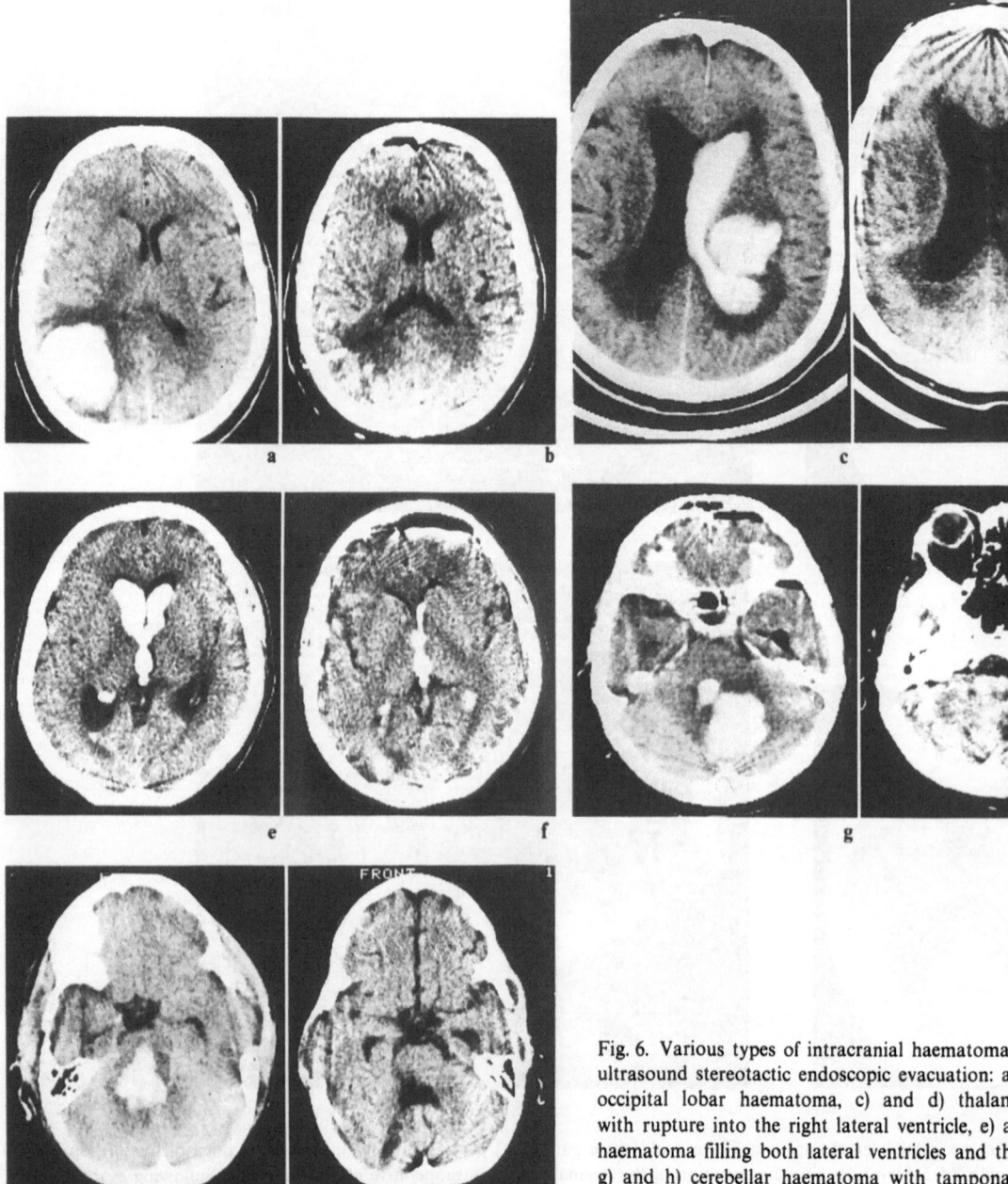

Fig. 6. Various types of intracranial haematoma before and after ultrasound stereotactic endoscopic evacuation: a) and b) parieto-occipital lobar haematoma, c) and d) thalamic haemorrhage with rupture into the right lateral ventricle, e) and f) ventricular haematoma filling both lateral ventricles and the third ventricle, g) and h) cerebellar haematoma with tamponade of the fourth ventricle, i) and j) traumatic brain stem haematoma

of the irrigation system which is used for the gentle movement of blood clots. Minor oozing from small vessels in the wall of the cavity can be coagulated by aid of the laser. Patients with ventricular haemorrhage either from a thalamic bleed or other sources may profit especially from an endoscopic procedure (see Figs. 6 c–f). The approach is either via a classical frontal burr hole or a parieto-occipital burr hole. Also in cases of cerebellar haematoma, the method has been used with success. In addition to evacuation of the haematoma itself, blood clots in the fourth ventricle may be washed in order to prevent the development of shunt-dependent hydrocephalus in these patients.

Ventricular Tumours

Tumours in the lateral ventricles or tumours in the third ventricle enlarging a foramen of Monro and bulging into the lateral ventricle can be approached with this endoscopic technique. The main indication to use an endoscope for biopsy and resection of the tumour is blockade of the foramina of Monro (tumours that do not lead to blockade of the CSF pathways can more easily be approached with a stereotactical biopsy needle alone for biopsy or evacuation of a cyst, see Fig. 7). The advantage of the endoscopic approach in these cases is a smaller approach through viable

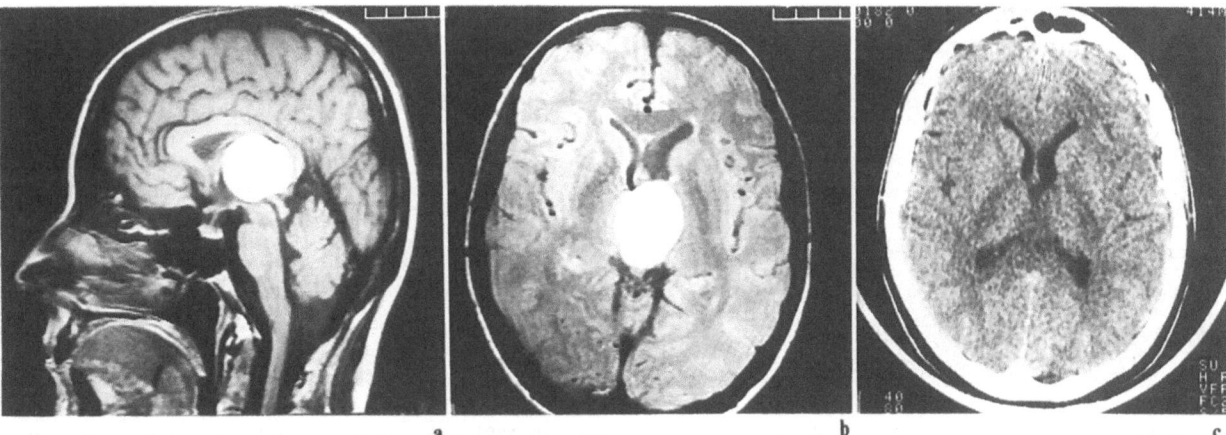

Fig. 7. Colloid cyst of the third ventricle a) and b) in the preoperative MRI image, c) control computerized tomography following ultrasound stereotactic punction under local anaesthesia and drainage of the cyst

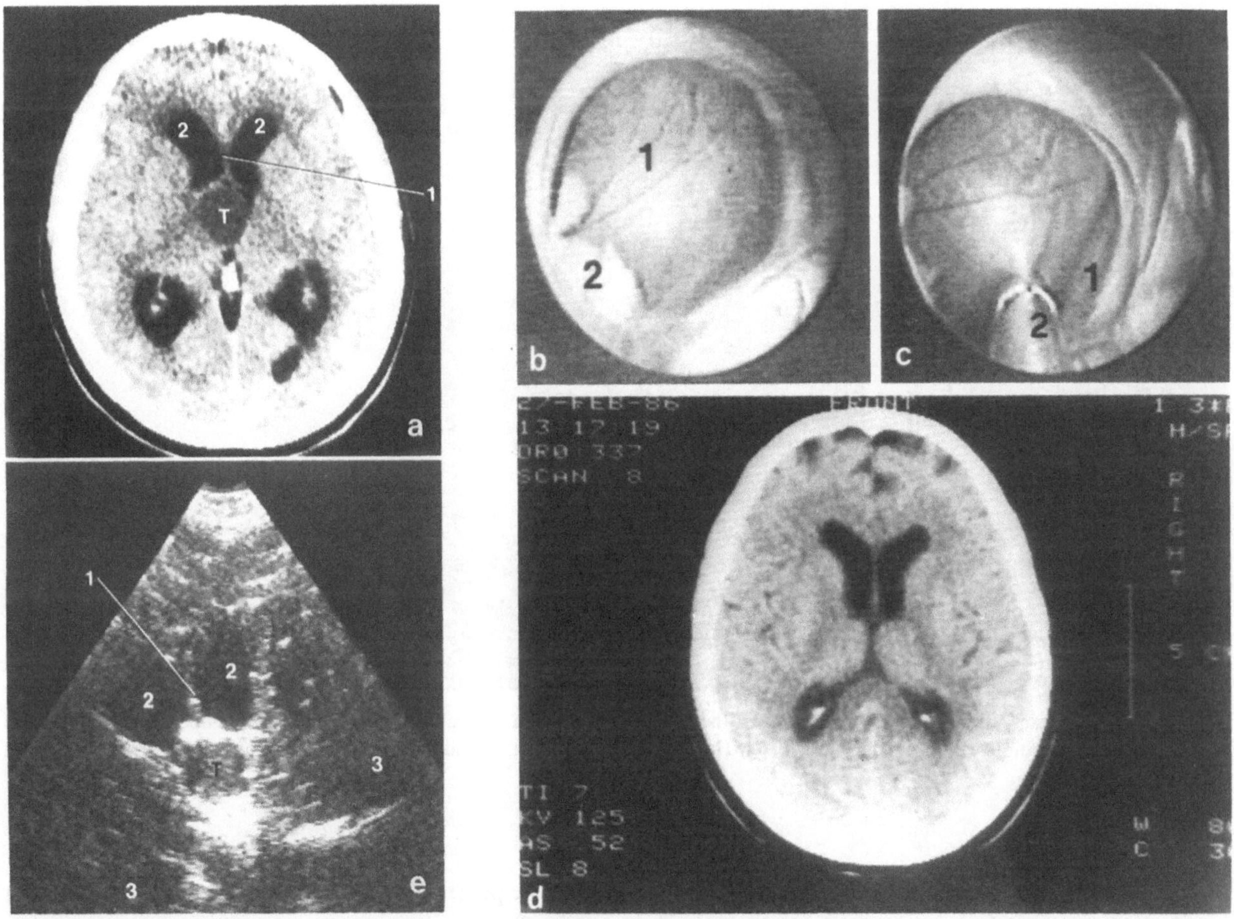

Fig. 8. Cystic craniopharyngioma in a 30 year old patient suffering from blockade of both foramina of Monro and chronic headache as well as attacks of vertigo and hemiparesis on the right side: a) pre-operative computerized tomogram, b) intra-operative images during ultrasound stereotactic endoscopic resection of the tumour: view into the right lateral ventricle towards the enlarged foramen of Monro on the right side; 1 vascularized membrane of the cystic tumour, bulging from the third ventricle through the foramen of Monro into the lateral ventricle, 2 choroid plexus; c) view on the surface of the tumour during its opening with a microbiopsy forceps (2) following laser coagulation of the tumour surface. 1 floor of the third ventricle. d) Control CT one month postoperatively. Normalization of the ventricular volume, the tumour is removed. e) Intra-operative ultrasound image, coronal section through a right frontal burr hole: 1 septum pellucidum, 2 lateral ventricle, 3 temporal lobe, T tumour

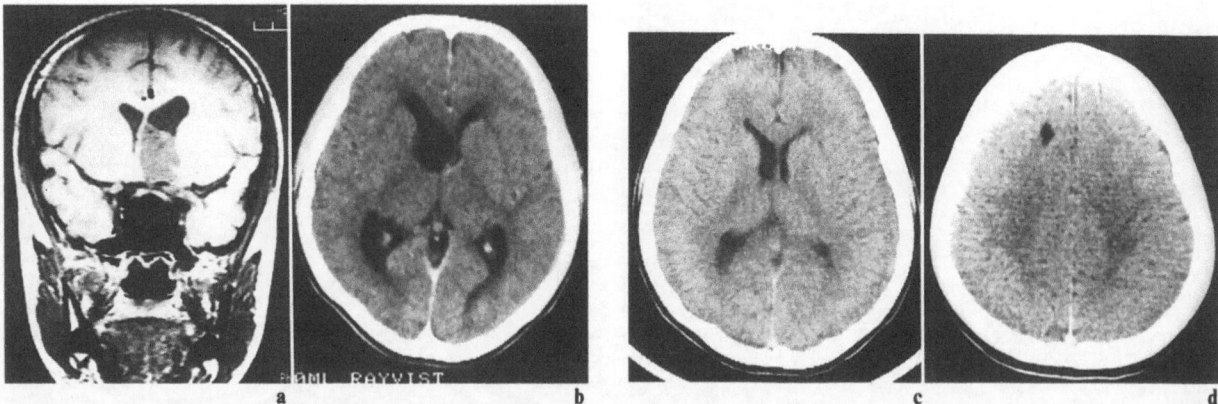

Fig. 9. Low grade astrocytoma in an 18 year old female patient: a) pre-operative MRI image, coronal section, b) pre-operative computerized tomogram, horizontal section: The tumour fills the third ventricle and reaches through the enlarged left foramen of Monro into the left lateral ventricle. Enlargement of the lateral ventricular system by blockade of both foramina of Monro, especially the left. c) Control computerized tomogram following ultrasound stereotactic endoscopic resection and CT guided stereotactic seed implantation and interstitial irradiation (Prof. Ostertag), d) surgical defect following endoscopic resection via left frontal trepanation

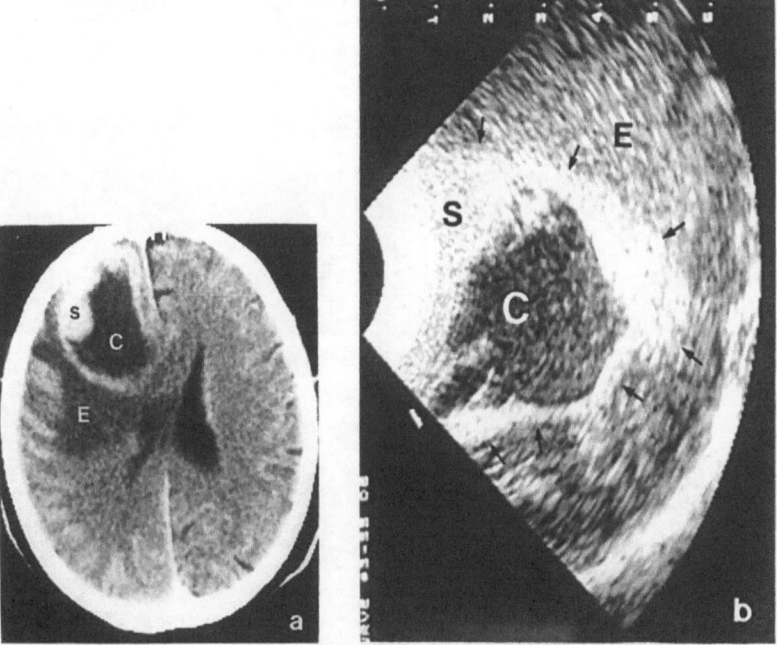

Fig. 10. Cystic metastasis from bronchial carcinoma in the left frontal lobe of a 67 year old male patient: a) pre-operative computerized tomogram, b) intra-operative ultrasound image with 5 MHz-probe, C tumour cyst, S solid tumour, E surrounding œdema, arrows: suspected border of tumour

cerebral tissue into the ventricular system. Tumours of the third ventricle are approached through the frontal horn via a frontal burr hole on the side of more prominent hydrocephalic enlargement. The surface of the tumour is then visualized as it bulges through the enlarged foramen of Monro. The surface of the tumour is coagulated with the aid of the laser. Biopsies can then be taken with biopsy forceps. In some cases, it

may be possible to extract the wall of a cystic tumour (Fig. 8).

In cases of solid tumours too large for radical removal, endoscopy may be used for size reduction to restore the CSF pathways. Thereafter, the residual tumour may be treated further by implantation of radio-active seeds (low grade astrocytomas, craniopharyngiomas) (Fig. 9).

Cystic Hemispheric Tumours

Cystic astrocytomas and cystic metastases may grow rapidly and lead to rather sudden decompensation of patients. Surgical decompression can be achieved by simple burr hole drainage; however, additional endoscopy allows one to take biopsies from the cyst wall under visual control. Morever, some shrinkage of the cyst wall may be achieved by irradiation with the laser. The residual solid tumour can thereafter be treated by percutaneous irradiation or implantation of a radio-active seed or by conventional neurosurgery.

References

1. Auer LM, Holzer P, Ascher PW, Heppner F (1988) Endoscopic neurosurgery. Acta Neurochir (Wien) 90: 1–14
2. Auer LM, van Velthoven V (1990) Intraoperative ultrasound (US) imaging. Comparison of pathomorphological findings in US and CT. Acta Neurochir (Wien) 104: 84–95
3. Auer LM, van Velthoven V (1990) Intraoperative ultrasound imaging in neurosurgery. Comparison with CT and MRI. Springer, Berlin Heidelberg New York
4. Berger MS (1985) Ultrasound guided stereotactic biopsy using the diasonics neuro-biopsy device for deep seated intracranial lesions. Presented at the Annual Meeting of the American Association of Neurological Surgeons, Atlanta, Georgia, April 21–25, 1985
5. Brown FD, Rachlin JR, Rubin JM, Fessler RG, Smith LJ, Schaible KL (1984) Ultrasound-guided stereotaxis. Neurosurgery 156, no 2: 162–164
6. Duthel R, Portafaix M (1986) Biopsies echoguidées des tumeurs cérébrales. Neurochirurgie 32: 547–552
7. Ostertag ChB (1987/88) Experimentelle Dosiseffekte als Grundlage der stereotaktischen interstitiellen Bestrahlung von Hirntumoren. Annales Universitatis Saraviensis Medicinae, Band 26, Nummer 1: 3–9
8. Tsutsumi Y, Andoh Y, Inoue N (1982) Ultrasound guided biopsy for deep-seated brain tumors. J Neurosurg 57: 164–167
9. van Velthoven V, Auer LM (1990) Practical application of intraoperative ultrasound imaging. Acta Neurochir (Wien) 105: 5–13

Correspondence: Prof Dr. L. M. Auer, Neurosurgical Clinic, Saarland University Medical School, D-6650 Homburg/Saar, Federal Republic of Germany.

Acta Neurochirurgica, Suppl. 54, 42–46 (1992)

Fenestration of Intraventricular Cysts Using a Flexible, Steerable Endoscope

S.K. Powers

Neurosurgical Oncology, Division of Neurosurgery, University of North Carolina, Chapel Hill, North Carolina, U.S.A.

Summary

Multiloculated hydrocephalus with multiple intraventricular septae due to meningitis associated with ventriculitis and other CSF containing intraventricular cysts can be treated by endoscopic fenestration. Seven patients with various CSF containing intraventricular cysts were treated using a flexible steerable endoscope and the argon laser. The experience using several currently available steerable endoscopes for treatment of this neurosurgical problem is reviewed. Emphasis is placed on the use of the laser for cyst fenestration. Successful decompression of the ventricular cyst(s) was accomplished in five cases with the endoscope alone. Craniotomy was required in two patients in order to complete cyst fenestration. It is the author's impression that laser assisted ventriculoscopy with steerable flexible endoscopes is an alternative and oftentimes superior method of treating CSF containing cysts within the lateral ventricles of hydrocephalic patients. Steerable flexible endoscopes designed specifically for neurosurgical use are needed.

Keywords: Argon laser; hydrocephalus; intraventricular cysts ventriculoscopy.

Introduction

The treatment of multiloculated hydrocephalus following cerebrospinal fluid (CSF) infections continues to be a difficult neurosurgical problem, and surgical treatments have included multiple shunt placements, use of multiperforated ventricular catheters, and craniotomy with fenestration of ventricular septae to convert multilocular to unilocular hydrocephalus[1,7,8]. Preliminary experience using a flexible, steerable endoscope and the argon laser to fenestrate intraventricular cysts associated with multiloculated hydrocephalus due to meningitis and ventriculitis was reported in 1986[4]. The use of intraventricular endoscopic techniques to fenestrate ventricular septae through a standard bur hole and small corticotomy have the advantage over craniotomy for fenestration in that the latter procedure tends to lose CSF leading to collapse of the ventricular walls during decompression.

The production of steerable, flexible small caliber fiberoptic endoscopes has greatly increased the navigational capabilities of the surgeon through the tracheobronchial tree, the gastrointestinal tract, and now the ventricular system of the brain. The flexible endoscopes consist of a small diameter fiberoptic bundle that is coupled to a lens system that is used by the surgeon for direct viewing or attached to a television camera for the surgeon to view on a television monitor. Illumination is provided by additional fiberoptic bundle(s) within the endoscope that carry intense white light from a light source into the surgical cavity. One or two hollow channels in the endoscope are utilized for the insertion of surgical instruments or laser fibers and for suction and irrigation. A steering mechanism near the viewing lens system allows the surgeon to manipulate the tip of the endoscope from side to side without manipulating the portions of the scope near the lens system.

In this report, I review my experience using various flexible endoscopes and the argon laser for fenestration of intraventricular cysts.

Materials and Methods

Case Material

Between February, 1985 and March, 1988 seven patients with various intraventricular cysts were treated using a flexible steerable endoscope and the argon laser. This series includes two males and five females ranging in age from four months to 23 years. Four children were treated for multiloculated ventricles with hydrocephalus secondary to previous intraventricular infection. One patient was treated for unilateral hydrocephalus resulting from previous intraventricular hemorrhage. One patient had an enlarging occipital porencephalic cyst resulting in obstructive hydrocephalus, and

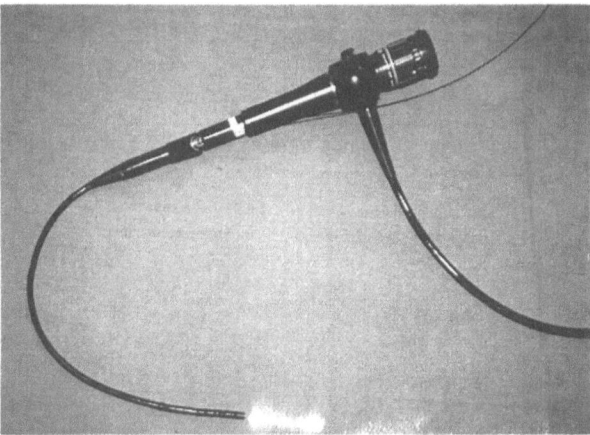

Fig. 1. Flexible steerable fiberscope (Model CHF-P10, Olympus Corporation) currently used for ventriculoscopy by the author. Optical laser fiber is seen exiting the treatment channel of the endoscope

another patient was treated for a large interhemispheric chronic subdural hematoma associated with massive congenital hydrocephalus. Two of the patients with multiloculated hydrocephalus have been previously reported[4].

The first four patients in this series were treated using a flexible pediatric bronchoscope (Model BF-P10; Olympus Corp. of America, New Hyde Park, New York) that had an outer diameter of 3.2 mm and a single treatment port measuring 2.0 mm in diameter. A 6.0 mm outer diameter, multiple-channeled flexible bronchoscope (Model BF-IT10, Olympus Corp.) was used in one of the children with multiloculated ventricles so that sapphire-tipped optical laser fibers could be used in an attempt to improve the cutting of the intraventricuar cyst membranes. This particular endoscope is equipped with two 2.6 mm diameter treatment ports, each capable of passing the 2.2 mm diameter sapphire-tipped laser fibers. Unfortunately, there was no advantage of the sapphire-tipped optical laser fiber over the bare-tipped fiber with the argon laser inside the ventricle. This will be discussed later in a case example. The last two patients were treated using a smaller caliber and shorter lengthed flexible choledochoscope (Model CHF-P10, Olympus Corp.) which is currently preferred by the author due to its size and greater ease of use for flexible ventriculoscopy (Fig. 1).

Technique

The endoscope is gas-sterilized along with all the laser fiberoptics and biopsy instruments for the endoscope on the day before operation. The television camera equipment is draped from its attachment point to the endoscope back to the television monitor. Under general anesthesia, all patients were placed in a prone position with the head supported on the cerebellar horseshoe headrest. A linear scalp incision and bur hole are made over the thinnest region of the occipital cortex as determined by the preoperative CT scan. The dura is opened and the lateral ventricle is initially entered with a 16 gauge cone ventricular needle. The needle tract is enlarged by gently rotating the hub of the needle or further incising the surrounding cortex to allow passage of the flexible endoscope.

A bent malleable retractor attached to the Yasargil retraction system is used to hold the endoscope stationary at the burr hole site. Navigation of the ventricular system is accomplished by vertical or lateral movements of the tip of the endoscope using the steering mechanism and by manually inserting and withdrawing the endoscope from the burr hole. Suction and irrigation with normal saline are performed by manipulating a three-way stopcock that is attached to the tubing that goes to the irrigation port of the endoscope. Instruments such as the laser optical fibers and biopsy forceps are inserted into a separate port into the hollow treatment channel of the endoscope.

After the cyst wall is identified, the laser fiberoptic is introduced through the treatment port and brought out to the end of the endoscope until the tip can be visualized on the television screen. The size of the laser beam at the tissue is then adjusted by varying the distance between the bare-tipped optical fiber and the tissue to be coagulated. By using a 1 to 2 mm diameter laser beam and approximately 3 to 4 watts of argon laser power for 0.5 to 2 seconds, the cyst wall is coagulated in a stepwise fashion. Small vessels within the wall can be seen to blanch during this maneuver. The laser power is then increased to between 7 and 14 Watts at the laser console and the spot diameter of the beam is decreased to about 0.5 mm in diameter. The laser energy is then sequentially applied for 0.5 second intervals to the tissue until perforation of the wall is seen. More recently, use of the laser fiber tip as a "hot knife" in direct contact with the tissue has been used for incision of cyst membranes using 7 to 10 Watts of laser energy delivered to the optical fiber. The sapphire-tipped fibers do not appear in my experience to have any advantage over bare-tipped laser fibers when used in a fluid filled ventricular cavity.

Usually, particulate matter within the CSF can be seen moving in or out of the cyst depending upon the pressure differential that exists between the cyst and the ventricular system. Multiple perforations can be made in the cyst wall circumferentially and then the central portion of the cyst wall removed either with the biopsy forceps or punctured through by advancing the endoscope. Likewise, a direct linear incision can be made in the cyst membrane using direct contact of the laser fiber and sweeping the laser fiber through the cyst by steering the endoscope during laser application. Venous bleeding from the small subependymal veins can be stopped by using low power densities of laser energy applied to coagulate, shrink, and thus occlude the vessel.

After the cysts are communicated with the ventricular system, the ventricle is irrigated through the endoscope with normal saline until all particulate matter and blood are removed and then the endoscope is withdrawn. The dura matter is left open and the scalp over the bur hole is closed. Usually, patients are followed in the intensive care unit overnight for possible complications.

Results

The treatment results in these seven patients are presented in Table 1. Five of the patients were improved by the endoscopy procedure alone with decompression of the cyst(s) and/or the ventricles after endoscopic laser fenestration (Fig. 2). Intraoperative complications were encountered in three of the patients, two of which required abandonment of the endoscopic procedere and the performance of a craniotomy to complete the fenestration. These 3 cases will be discussed in further detail.

One of these patients was a 23 year old woman with a right occipital porencephalic cyst with hydrocephalus due to upper brainstem compression from an extension of the cyst through the tentorial incisura into the cisterna ambiens. The cyst was endoscopically opened into the frontal and temporal horn of the right lateral ventricle without difficulty; however, during

Table 1. *Fenestration of Intraventricular Cysts with Argon Laser Through Flexible Endoscopy*

| | | Case Summaries | | |
Patient	Diagnosis	Procedure	Complications	Results
(1) 1 YO Male	Multiloculated ventricles	Fenestration of intraventricular cysts	—O—	Improved
(2) 7 MOS Male	Multiloculated ventricles	Fenestration of intraventricular cysts	—O—	Improved
(3) 23 YO Female	Occipital porencephalic cyst, right	Fenestration of cyst with right lateral ventricle	Venous bleeding, craniotomy	Improved
(4) 4 MOS Female	Multiloculated ventricles	Fenestration of intraventricular cysts	Cyst walls too thick, craniotomy	Improved
(5) 1 YR, 8 MOS female	Congenital hydrocephalus chronic interhemispheric SDH	Fenestration and drainage of SDH	—O—	Improved
(6) 1 YR, 4 MOS female	Multilocated ventricles	Fenestration of intraventricular cysts	—O—	Improved
(7) 4 Mos. Female	Unilateral obstructive hydrocephalus, left	Performation of septum pellucidum	Intraventricular/ septal hematoma	Improved

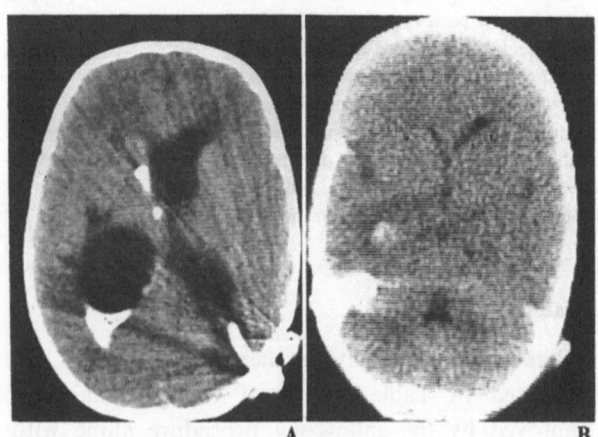

Fig. 2. Patient No. 2. A) Pre-operative metrizamide CT scan. CSF-containing cysts are present at the foramen of Monro and in the atrium of the left lateral ventricle. B) Post-operative non-contrast CT scan (2 days after surgery). The ventricles are decompressed due to communication of the cysts with the lateral ventricles. A small hematoma is present in the temporal horn of the left lateral ventricle

exploration of the cyst in the region of the dorsal mesencephalon, hemorrhage was encountered below the tentorial edge, the source of which could not be visualized through the endoscope. The endoscope was, therefore, withdrawn and a 4 cm diameter occipital craniotomy was made. The corticotomy was enlarged and the cyst explored. By the time microscopic visualization of the area was

obtained the haemorrhage had stopped. The cyst was then fenestrated under microscopic visualization into the ambient cistern. Communication between the right lateral ventricle, the cyst, and the basilar cisterns was verified by placing Indigo carmine into the frontal horn of the right lateral ventricle and watching the dye pass up from around the brainstem through the fenestration made in the medical aspect of the cyst floor from the ambient cistern. The patient had an uneventful recovery and the lateral ventricles and cyst were decompressed on the postoperative CT scan.

Craniotomy was needed also in a four month old female born with a thoracolumber myelomeningocele and multiloculated hydrocephalus due to previous *Providencia sp.* ventriculitis in whom the laser was unable to effectively open the intraventricular cysts. This was the only patient in which the larger flexible bronchoscope, measuring 6 mm in diameter, was used in order to take advantage of the multiple treatment channels within the scope that could allow the passage of the 2.2 mm diameter sapphire-tipped laser fibers that theoretically should improve the focusing of laser energy and thus facilitate cutting and vaporization with the argon laser. However, this particular case illustrates the fact that sapphire-tipped fibers in CSF or any other type of fluid medium lose their laser focusing abilities and prove to be less beneficial than bare-tipped laser fibers for incising the cyst membranes.

In this patient, due to difficulty with cutting cyst membranes using the laser fibers and clouding of the CSF from slow hemorrhage from the laser incisions, the endoscopic procedure was abandoned in favor of open craniotomy which was then performed. Interestingly, there was extensive neovascularization and gliosis of the membranes within the ventricular system and a total of seven cysts were communicated in order to create a single ventricular cavity.

The third complication is that of an intraventricular, intraseptal (septum pellucidum) hemorrhage during fenestration of the septum pellucidum in a four month old female with unilateral hydrocephalus due to an adhesion at the Foramen of Moro. Intraoperative ultrasonography through the anterior fontanel was used to guide the endoscopic approach. During mechanical removal of a 1 cm perforation within the septum pellucidum subependymal venous bleeding was encountered. A clot formed in the region of the fenestration and attempts at clot removal were met with recurrent hemorrhage. After removal of the endoscope a ventricular catheter was placed into the ventricle allowing drainage of blood tinged CSF for three days. After ventriculostomy, the clot resolved and the fenestration remained patent for six months. However, the patient's right lateral ventricle again became enlarged and was treated nine months later with a unilateral ventriculoperitoneal shunt.

Discussion

Although endoscopic cauterization of the choroid plexus was first attempted by the urologist L'Espinasse in 1910[6], Mixter was the first to report use of the operating urethroscope to perform third ventriculoscopy in 1923[2]. There have been several reports describing the use of rigid fiberoptic endoscopes for performing third ventriculostomy and performing biopsy and aspiration of colloid cysts of the third ventricle during the past fifteen years[3,9]. In spite of the increased enthusiasm for the use of ventriculoscopy by neurosurgeons using small diameter rigid fiberoptic ventriculoscopes during the past several years, the technique has not yet enjoyed widespread popularity. One of the reasons for this may be that the rigid ventriculoscopes do not permit exploration of the entire ventricular system, particularly in patients with only modest enlargement of the ventricles without requiring an excessive amount of manipulation of the endoscope at its cortical entrance into the brain.

Likewise due to the design of most rigid scopes, often surgical manipulations cannot be performed under direct vision[9].

Steerable flexible endoscopes have improved the ability of the surgeon to maneuver during endoscopy. Added to this is the recent development of laser technology and, in particular, visible wavelength lasers such as the argon laser which can be directed down fine optical fibers that can be inserted through the treatment channel within the flexible endoscope[5]. The laser energy can be delivered at the end of the endoscope under direct vision of the surgeon. With the argon laser the aiming beam is an attenuated treatment beam. This allows the surgeon to direct the laser thermal effect exactly to the area outlined by the aiming beam from the tip of the fiber. The argon laser can be fired through clear fluid such as CSF which enables one to operate in the ventricular system without significant attenuation of the laser energy. The spot diameter to the laser beam can be increased or decreased by moving the optical fiber toward or away from the tissue to be treated. By making adjustments in the spot diameter and power output of the laser the surgeon can select among coagulation, vaporization, and incision.

Although there does not currently exist a flexible endoscope designed for intraventricular endoscopic procedures, the use of currently existing small caliber endoscopes such as the choledochoscope can be used safely in their present state for diagnostic evaluation and for surgical manipulation of relatively avascular lesions in the lateral and third ventricles of hydrocephalic patients.

There is, however, an obvious need for the development of one or more flexible, steerable endoscopes for specific use in cerebral ventricular endoscopy. An endoscope capable of continuous as well as intermittent suction and irrigation from its tip that can be controlled by foot pedal(s) or automatically is preferable in order to free the surgeon's hands which are involved in manipulating the endoscope and instrumentation through its channels. The development of an "approximating" bipolar-tipped forceps and other microendoscopic instruments designed for neurosurgical use as alternative instruments to the use of optical laser fibers are essential in order to further microendoneurosurgery for use in diagnosis and treatment of disorders involving the cerebral ventricles.

We are optimistic that these developments in endoscope and instrument design are close at hand.

References

1. Albanese V, Tomasello F, Sampaolo S (1981) Multiloculated hydrocephalus in infants. Neurosurgery 8: 641–646
2. Mixter WJ (1928) Ventriculoscopy with puncture of the floor of the third ventricle. Boston Med Surg J 188: 277–278
3. Powell MP, Torrens MJ, Phil M, Thomson JLG, Horgan JG (1983) Isodense colloid cysts of the third ventricle. A diagnostic and therapeutic problem resolved by ventriculoscopy. Neurosurgery 13: 234–237
4. Powers SK (1986) Fenestration of intraventricular cysts using a flexible, steerable endoscope and the argon laser. Neurosurgery 18: 637–641
5. Powers SK, Edwards MSB, Boggan JE, Pitts LH, Gutin PH, Hosobuchi Y, Adams JE, Wilson CB (1984) Use of the argon surgical laser in neurosurgery. J Neurosurg 60: 523–530
6. Pudenz RH (1981) The surgical treatment of hydrocephalus – an historical review. Surg Neurol 15: 15–26
7. Rhoton AL, Gomez MR (1972) Conversion of multilocular hydrocephalus to unilocular. J Neurosurg 36: 348–350
8. Schultz P, Leads NE (1973) Intraventricular septations complicating neonatal meningitis. J Neurosurg 38: 620–626
9. Vries JK (1978) An endoscopic technique for third ventriculostomy. Surg Neurol 9: 165–168

Correspondence: Prof. S.K. Powers, M.D., F.A.C.S., Director of Neurosurgical Oncology, Division of Neurosurgery, University of North Carolina, Chapel Hill, North Carolina, U.S.A.

Acta Neurochirurgica, Suppl. 54, 47–52 (1992)
© by Springer-Verlag 1992

Application of Superfine Fiberscope for Endovasculoscopy, Ventriculoscopy, and Myeloscopy

K. Yamakawa, T. Kondo, M. Yoshioka, and **K. Takakura**

Department of Neurosurgery, National Medical Cantor of Tokyo and University of Tokyo Hospital, Tokyo, Japan

Summary

In the past three years, we have used a superfine fiberscope for endovasculoscopy, ventriculoscopy, and myeloscopy. Flexible superfine fiberscope, 0.75 mm in outer diameter, could visualize various intravascular findings.

In in-vivo canine experiment, sequential changes of thrombus produced by endothelial abrasion by needle or balloon in the canine carotid artery can be clearly seen. And thrombolysis by focal arterial injection of tissue plasminogen activator was sequentially observed.

In the *clinical study*, we could evaluate the stenotic lesions of the subclavian and vertebral arteries before and after balloon angioplasty.

During ventriculoscopy, a 2 mm of mini-caliber fiberscope was introduced under ultrasound monitoring, and provided clear visualization of intraventricular tumours.

In myeloscopy, draining veins of arteriovenous malformations and nerve roots of the cauda equina could be clearly seen by flexible fiberscope.

From these results, it can be said that the superfine fiberscope provides clear and useful visualizations of the interior of vessels, ventricles, and the intrathecal area of the spinal canal. The new applications of this superfine fiberscope for minimally invasive neurosurgery may bring about a marked improvement of therapeutic results.

Keywords: Superfine fiberscope; endovasculoscopy; ventriculoscopy; myeloscopy.

Introduction

Direct observation of the inner cavities of the human body is very important for correct lesional evaluation and diagnosis. In the field of neurosurgery, many practical interventions using the endoscope for ventriculoscopy[4,6,7,8,10,11,14,15] and myeloscopy[9,12,13] have been performed for the past 80 years. In spite of this prolonged trail of endoscopy, endoscopic neurosurgery did not spread widely until recent days because this approach could not offer any real advantages for neurosurgical treatment in comparison with current surgery. But, with the advent of high resolution fiberoptics in the recent several years, our understanding of endoscopic neurosurgery has been revolutionized.

In the past 3 years, we have used superfine fiberscope for endovasculoscopy, ventriculoscopy, and myeloscopy. In this paper, this new superfine fiberscope is introduced, and its clinical usefulness is discussed.

Methods

Construction and Components of the Superfine Fiberscope and Fiber Imaging System

The flexible superfine fiberscope is composed of image guide, which contains 4200 image bundle of silica fibers, and light guide of multi-component glass fibers. The image guide and light guide are confined in a Teflon catheter (Fig. 1). These silica imaging fibers are superior to the conventional glass fiber in respect of fineness, durability, and cost.

A superfine fiberscope can be designed in many ways according to the various medical uses. For endovasculoscopy, we usually use a flexible fiber scope, 150 cm long and 0.75 mm in outer diameter. We designed the mini-caliber fiberscope, 15 cm long, 2 mm in outer diameter contained in stainless steel for ventriculoscopy with an installed irrigation channel. Superfine fiberscope generally has a 70 degrees angle of view, and its depth of observation is 1 mm to 20 mm. Unlike conventional fiberscopes, these thin fiberscopes have no eyepiece, and more than one person can get a fine optical image from the color monitor. An integral ocular optical system, TV-camera, light source, color monitor, and video processor are built into the fiber-imaging system. A 35 mm camera can be fitted to the fiber-imaging system, and it is possible to take a photograph whenever it is appropriate (Fig. 2). These superfine fiberscopes and fiber imaging systems were developed by Medical Science Co., Ltd and are now on the market by Mizuho medical Co., Ltd.

Application of Superfine Fiberscope

Endovasculoscopy

The superfine fiberscope, 150 cm long and 0.75 mm in outer diameter, was used for endovasculoscopy. For observation of the

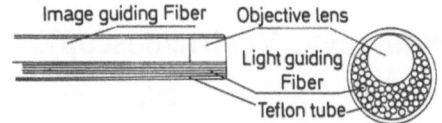

Fig. 1. Construction of the superfine fiberscope

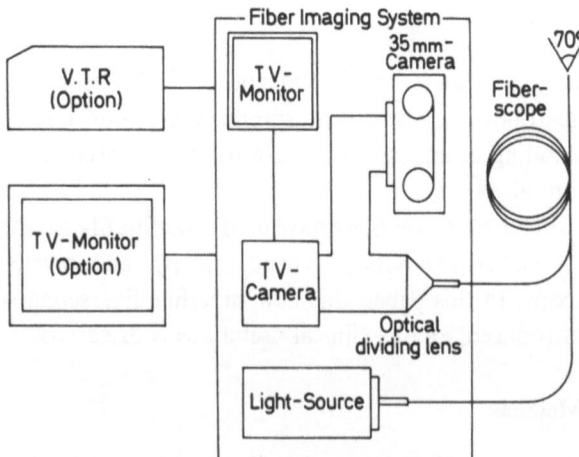

Fig. 2. Fiberscope and fiber imaging system

interior of the vessel, a 7 French occlusion balloon equipped with double lumen was used with continuous flush of physiological solution. The superfine fiberscope can easily be introduced through the balloon catheter using the Seldinger technique.

Intraluminal visualization of vessels had been studied using canines prior to its clinical application. In the canine carotid artery, thrombus could be produced after endothelial injury using balloon and needle. Development of thrombus formation and thrombolysis after arterial infusion of tissue plasminogen activator (250–500 ug) were endoscopically observed.

In in-vivo human studies, the obstructive and stenotic site of the subclavian artery, orifice of the vertebral artery, jugular vein, and bifurcation of the common carotid artery were endoscopically observed. In the cases with stenosis at the left subclavian artery and the orifice of the vertebral artery, endovasculoscopy was performed before and after balloon angioplasty.

Ventriculoscopy

The mini-caliber fiberscope as described above was used for ventriculoscopy. At first, we had experience of intraventricular observation of hydrocephalus due to intraventricular haemorrhage in an immature infant. When the ventriculo-peritoneal shunt was performed, a superfine fiberscope was introduced through the burr hole of the frontal bone. Although, clear visualizations of intraventricular structures were obtained, we could not obtain

precise anatomical orientation. After this experience, it was thought that another parameter for guidance such as ultrasound or stereotaxy was necessary for correct endoscopic observation. In the present study, an Ultrasound camera and a 5.0 MHz micro intraoperative probe (Aloka Co., Ltd) were used for guiding the minicaliber fiberscope.

Myeloscopy

For myeloscopy, the same flexible fiberscope as for endovasculoscopy was used. Flexible fiberscope was introduced into the lumber subarachnoid space in a similar manner as lumber spinal drainage. Various endoscopic observations were obtained by pushing the fiberscope upward in the subarachnoid space. In the present study, a case with spinal arteriovenous malformation was inspected before surgery.

Results

Endovasculoscopy

In in-vivo canine studies, thrombus produced by endothelial abrasion by needle or balloon is clearly observed. The red thrombus, thrombus covered with whitish fibrin net, some folds and haemorrhage of intraluminal wall were nicely seen (Fig. 3). Sequential changes of thrombolysis were also observed after arterial infusion of tissue plasminogen activator. In in vivo human studies, we successfully observed obstructive sites of the extra-cranial vessels such as the internal carotid, subclavian and vertebral arteries, and the jugular vein. In the cases with "pulseless" disease or cerebral infarction, arterial stenosis of the left subclavian artery or the orifice of the vertebral artery were endoscopically inspected before and after balloon angioplasty. In these cases, intraluminal visualizations of the stenotic sites seemed to reveal an atheromatous plaque, the surface of which appears yellowish, and relatively smooth (Fig. 4). In the case of a partial occlusion of the jugular vein due to a jugular foramen neurinoma, although the stagnation and slow stream of the blood were clearly seen at the occluding site, tumour did not protrude into venous lumen.

Ventriculoscopy

Figure 5 shows an intraventricular tumour observed by a 2 mm ventriculoscope. The light gray tumour and thin vessel on its surface can be clearly seen (Fig. 5). The vessels of the ventricular wall and choroid plexus near the foramen of Monro are also clearly seen. Intra-operative use of ultrasound was very helpful for guiding the mini-caliber fiberscope to the desired region.

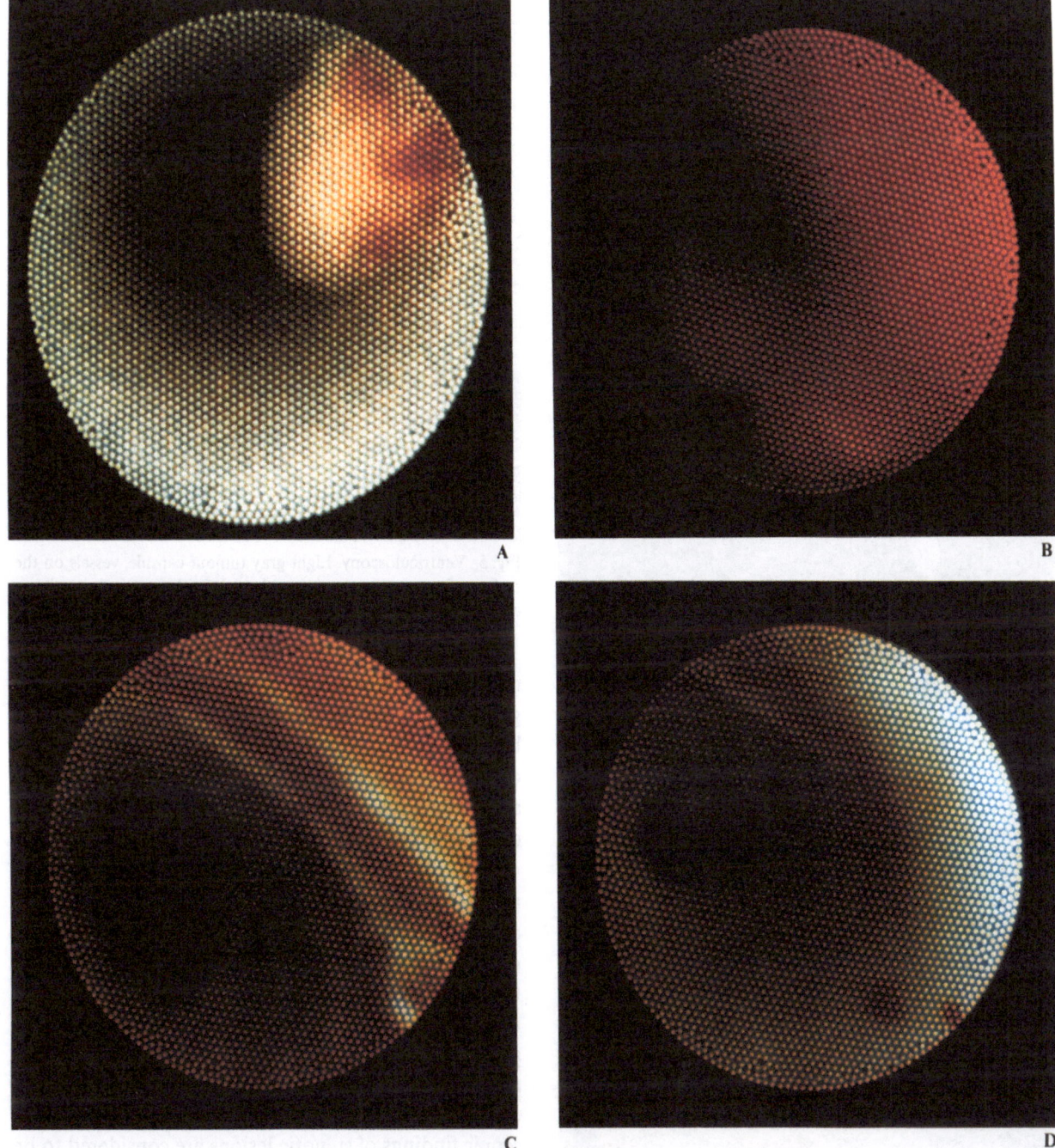

Fig. 3. Thrombus in canine carotid artery. A) Focal thrombus produced after endothelial abrasion by needle: Thrombus protrudes from the intraluminal wall. The surface of thrombus appears to be whitish and partially red. B), C), D) Red thrombus (B), irregular surface (C), and haemorrhage (D) of intraluminal wall of carotid artery are observed after endothelial abrasion with balloon

Myeloscopy

Figure 6 is the picture of a nerve root of the cauda equina, dura mater, and the distended draining vein of an arteriovenous malformation observed by myeloscopy. Thin vessels on the nerve root were also visible. Also other findings within the subarachnoid space near the conus medullaris can be obtained by upward introduction of the fiberscope.

Discussion

Direct endoscopic observation of the interior of the human body can provide three dimensional views, and is very important for making the correct diagnosis and instituting appropriate treatment. Since endoscopy was introduced for ventriculoscopy in 1910 in the neurosurgical field, the endoscope has been mainly used for observation of the interior of the ventricle for the treatment of hydrocephalus such as coagulation

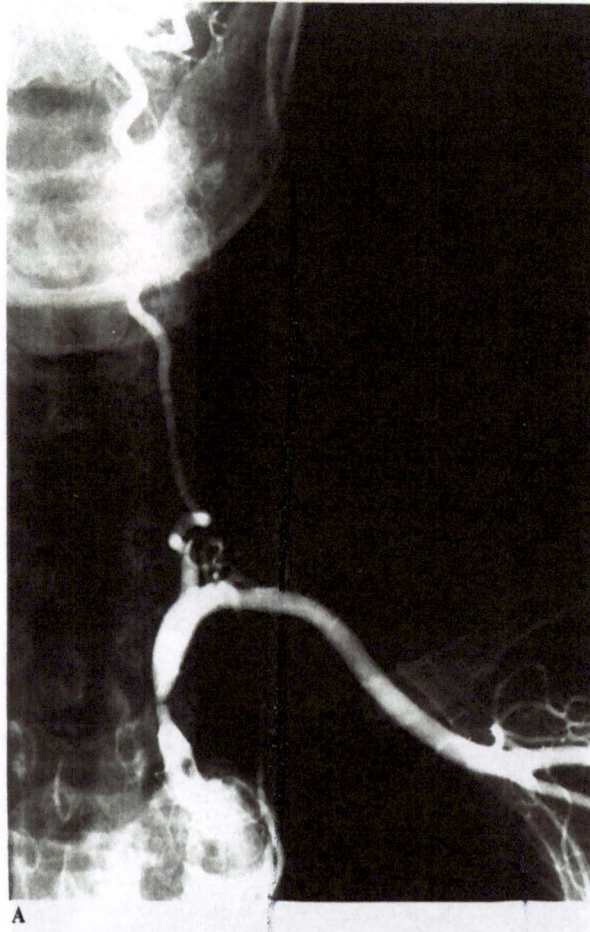

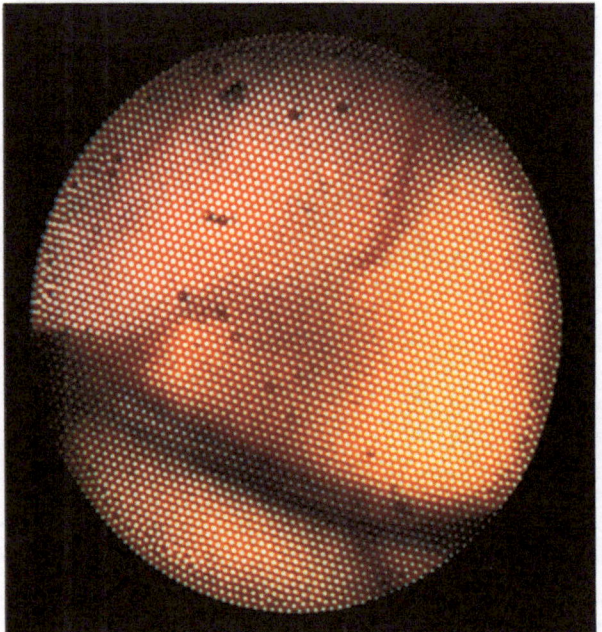

Fig. 5. Ventriculoscopy. Light gray tumour capsule, vessels on the surface of tumour, and vein on the yellowish ventricular wall are clearly seen

of the choroid plexus[4,6,7,14,15]. But endoscopic surgery has not widely spread in the neurosurgical field because of various problems. A new era has been opened in the neurosurgical field with the recent advent of a superfine fiberscope and the development of other technology using new instruments.

A flexible superfine fiberscope has been already applied to intraluminal observation of the coronary arteries[5,18,20] and some peripheral vessels[16,21], and its usefulness for percutaneous transluminal angioplasty[20] has been well documented. In the present experimental and clinical studies, several endoscopic findings of occlusive lesions of the vessels such as thrombus, atheroma, fibrin-platelet thrombi, and irregular surface of the vessel wall were nicely observed. These endoscopic findings of stenotic lesions are considered to be useful, since it provides some important information necessary for interventional procedures such as angioplasty, thrombolytic therapy, and/or medications. The use of this safe and useful endovasculoscope is strongly advocated for treatment of cerebrovascular disease due to stenosis of the extra cranial portions of the carotid, vertebral and subclavian arteries. Technical problems in endovasculoscopy are: 1) proximal ballon occlusion of arteries and saline flushing could not exclude the reflux blood sufficiently when the arterial stenosis was not so severe. 2) superfine fiberscope could not be placed to get a view of the stenotic site when the tip of balloon catheter did not faced desired portion

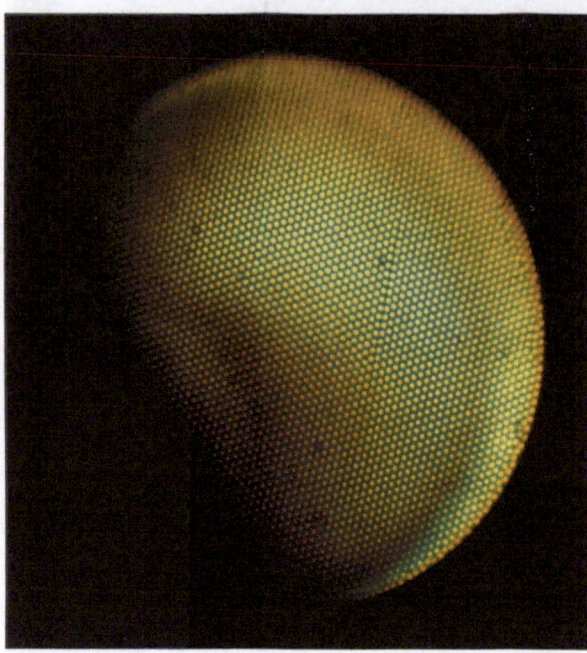

Fig. 4. Atheromatous plaque in stenotic pulseless disease: Brachial angiography showed 85% stenosis in the left subclavian artery (A). Using Seldinger's technique, superfine fiberscope catches a view of atheromatous plaque, the surface of which appears to be relatively flat and yellowish (B)

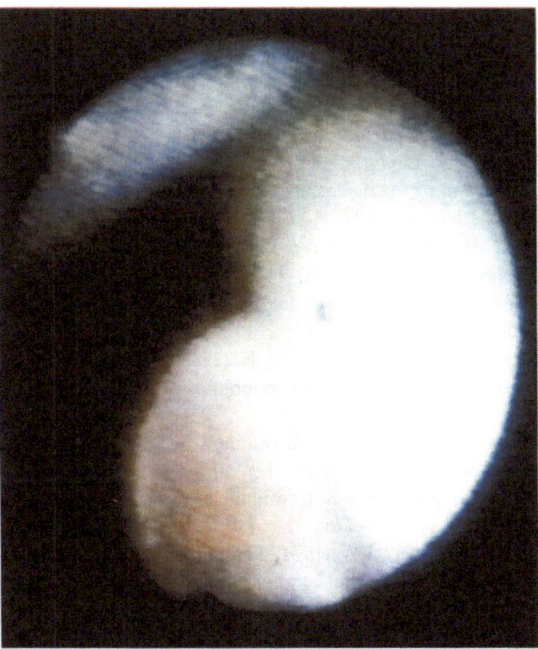

Fig. 6. Myeloscopy. Distended draining vein of a spinal arteriovenous malformation and nerve roots of cauda equina, and dura mater are observed

of artery. Then, another distal occlusion ballon for the prevention of reflux blood and development of superfine fiberscope which has angulation system might be necessary for detailed observation.

During ventriculoscopy, we can obtain clear visualization of tumour, vessels, ventricular wall, and choroid plexus using the 2 mm rigid fiberscope under ultrasound monitoring. A thin and flexible fiberscope certainly is not dangerous for the brain and neural tissue, however, it is difficult to detect by ultrasound. Furthermore, for endoscopic neurosurgery such as tumour biopsy or evacuation of haematoma[1,2], a rigid fiberscope is thought to be more applicable than a flexible fiberscope. Recently the usefulness of ultrasound-guided stereotaxic biopsy has been reported[3,17,19]. In our study, ultrasounds monitoring was thought to be helpful for guiding the superfine fiberscope to the target region. By using this ultrasounds guided endoscopy, intracerebral endoscopic surgery will be possible in near future.

In myeloscopy, it has definite merit for the establishment of the diagnosis of various kinds of diseases presenting with low back pain syndrome[12] and some myelopathy[9], which is difficult to evaluate by myelography, selective spinal angiography, and MRI. We have only two clinical experiences with myeloscopy. One is the case of a spinal AVM as we described above. In the other case with intractable pain due to end-stage

lung cancer, spinal cord and nerve roots were clearly identified before percutaneous upper-cervical cordotomy. It is now quite easy to see the subarachnoid space in the spinal canal using the flexible thin fiberscope. However, the visible field is restricted because of its thinness and flexibility. Flexible thin fiberscopes equipped with an angulation system and used under fluoroscopic guidance are thought to be necessary for more a definite diagnosis.

Acknowledgments

We are very grateful to R. Matsuura for her expert technical assistance for the canine studies, and to K. Choshoji, H. Tominaga, and Y. Maekawa for their researches of superfine fiberscope in Medical Sciences Co., Ltd.

References

1. Auer LM, Holzer P, Ascher PW, Heppner F (1988) Endoscopic Neurosurgery. Acta Neurochir (Wien) 90: 1–14
2. Auer LM, Deinsberger W, Niederkorn K, Gell G, Kleinert R, Schneider G, Holzer P, Bone G, Mokry M, Korner E, Kleinert G, Hanusch S (1989) Endoscopic surgery versus medical treatment for spontaneous intracerebral hematoma: a randomized study. J Neurosurg 70: 530–535
3. Berger MS (1986) Ultrasound-guided stereotaxic biopsy using a new apparatus. J Neurosurg 65: 550–554
4. Dandy, WE (1922) Cerebral ventriculoscopy. Bull Hopkins Hosp 33: 189–190
5. Forrester JS, Grundfest W, Litvack F (1989) Fiberoptic angioscopy in acute coronary syndromes. Coronary 6: 13–22

6. Griffith HB (1975) Technique of fontanelle and persutural ventriculoscopy and endoscopic ventricular surgery in infants. Child' brain 1: 359–363

7. Griffith HB (1987) Endoneurosurgery: Endoscopic intracranial surgery. In: Advances in neurosurgery, vol 14. Springer, Berlin Heidelberg New York, pp 2–24

8. Kleinhaus S, German R, Sheran M, Shapiro K, Boley SJ (1982) A role for endoscopy in the placement of ventriculoperitoneal shunts. Surg Neurol 18: 179–180

9. Miyamoto S, Kikuchi H, Nagata I, Yamagata S, Akiyama Y, Kaneko T, Ito K (1989) The development of spinal endoscope using a flexible optic fiber. Brain and Nerve 41: 1233–1238

10. Ogata M, Ishikawa T, Horide R, Watanabe M, Matsumura H (1965) Encephaloscope. J Neurosurg 22: 288–291

11. Oka K, Ohta T, Kibe M, Tomonaga M (1990) A new neurosurgical ventriculoscope-technical note. Neuro Med Chir 30: 77–79

12. Ooi Y, Satoh Y, Hirose K, Mikanagi K, Morisaki N (1977) Myeloscopy. Int Orthop 1: 107–111, 6: 881–894

13. Poll JL (1942) Myeloscopy. Intraspinal endoscopy. Surgery 11: 169–182

14. Putnam TJ (1934) Treatment of hydrocephalus by endoscopic coagulation of the choroid plexus. Description of a new instrument and preliminary report of results. New Eng J Med 210: 1373–1376

15. Scarff JE (1952) Non obstructive hydrocephalus. Treatment by endoscopic cauterization of the choroid plexus. J Neurosurg 9: 164–176

16. Seeger JM, Abela GS (1986) Angioscopy as an adjunct to arterial reconstructive surgery: A preliminary report. J Vasc Surg 4: 315–320

17. Sjolander U, Lindgren PG, Hugosson R (1983) Ultrasound sector scanning for the localization and biopsy of intracerebral lesions. J Neurosurg 58: 7–10

18. Spears JR, Marais J, Serur J, Pomerantzeff O, Geyer RP, Sipzener RS, Weintraub R, Thurer R, Paulin S, Gerstin R, Grossman W (1983) In vivo coronary angioscopy. J Am Coll Cardiol 1: 1311–1314

19. Tsutsumi Y, Andoh Y, Inoue N (1982) Ultrasound-guided biopsy for deep seated brain tumours. J Neurosurg 57: 164–167

20. Uchida Y, Hasegawa, K, Kawamura K, Shibuya I (1989) Angioscopic observation of the coronary luminal changes induced by percutaneous transluminal coronary angioplasty. Am Heart J 117: 769–776

21. Vollmar JF, Storz LW (1974) Vascular endoscopy possibilities and limits of its clinical application. Surg Clin North Am 54(1): 111–122

Correspondence: Kenta Yamakawa, M.D., Department of Neurosurgery, National Medical Center, 1-21-1, Toyamacho, Shinjuku-ku, Tokyo, Japan.

Acta Neurochirurgica, Suppl. 54, 53–58 (1992)
© by Springer-Verlag 1992

Percutaneous Endoscopic Laser Discectomy (PELD)
A New Surgical Technique for Non-sequestrated Lumbar Discs

H.M. Mayer, M. Brock, H.-P. Berlien*, and B. Weber*

Neurochirurgische Klinik, Universitätsklinikum Steglitz, Freie Universität Berlin,
Federal Republic of Germany
*LASER Medizin Zentrum Berlin GmbH, Berlin, Federal Republic of Germany

Summary

Basic features and techniques of percutaneous endoscopic laser discectomy are described and the results in 6 patients reported. Indications are: discogenic radicular symptoms, caused by disc protrusions, which do not respond to conservative treatment. Contra-indications are: major neurological deficit, segmental instability and spondylolisthesis, extruded disc prolapse, narrow spinal canal or lateral recess.

Keywords: Discectomy; laser; percutaneous operation; endoscopic operation.

Introduction

"Percutaneous nucleotomy" was deviced by Hijikata in 1975[6]. In recent years this procedure has been modified and considerably improved by various authors[7,8,11,12]. The different techniques are encompassed by the term "Percutaneous Lumbar Discectomy", and are described in detail elsewere[9]. The surgical goal of percutaneous discectomy is the selective removal of the herniated portions of nucleus pulposus in the dorsal third of the intervertebral space. This can be achieved through a postero-lateral approach from one or both sides, using rigid and flexible instruments under intermittent or permanent endoscopic control. Based on their experience with more than 100 cases of Percutaneous Endoscopic Discectomy, the authors describe a new technique which involves the *endoscopic* use of a Nd:YAG-LASER in the lumbar disc space.

Patient Selection

Clinical criteria (Table 1)

Percutaneous Endoscopic (LASER) Discectomy (PELD) is indicated in patients with discogenic radicular symptoms which do not respond to conservative treatment. It is a surgical method and is used as an alternative to other surgical procedures. Thus, indication for PELD includes the indication for lumbar microdiscectomy. However, since the decompression of the nerve root cannot be observed directly, PELD should be restricted to patients with minor or only slowly progressive neurological deficit (Table 1).

Morphologic criteria (Table 1)

The term "herniated lumbar disc" describes a pathomorphological entity which includes a large number of different "herniation forms" ranging from a bulging annulus to a sequestrated disc prolapse with large epidural fragments of nucleus pulposus. However, modern imaging techniques are usually able to provide a clear picture of the in-vivo pathomorphology of the intervertebral disc. The diagnostic efforts should result in a description of the morphology of the nucleus pulposus and annulus fibrosus as well as in the description of the topographic relationships of the disc herniation to intraspinal, intraforaminal or extraspinal structures. This can only be achieved by applying the whole spectrum of diagnostic aids such as Magnetic Resonance Imaging, spinal CT-scan in prone or supine position, Discography, Digital Sub-

Table 1. *Indications for Percutaneous Endoscopic LASER Discectomy*

Clinical aspect	Morphological aspect
Discogenic radicular symptoms:	Protrusion
– Sciatica	Subligamentous prolapse
– Sensory disturbance	– at the level of disc space
– Paresis (IV°–III°)	– < 1/3 of sagittal diameter
– Reflex-differences	of spinal canal
Unsuccessful conservative treatment L 1–S 1	

traction Discography and/or post-discographic CT-scan (Disco-CT). In cases of "stress-dependent" soft-disc herniations, stress-myelography should be performed.

Percutaneous Endoscopic LASER Discectomy (PELD) is indicated in symptomatic lumbar disc protrusions (annulus fibrosus intact) and in cases of a small subligamentous prolapse. Subligamentous prolapse should be at the level of the disc space without signs of cranial or caudal extension in the spinal canal. For technical reasons (size and working range of instruments), the prolapse should occupy less than 1/3 of the sagittal diameter of the spinal canal. There should be no signs of sequestration.

Contraindications (Table 2)

Clinical contra-indications include severe or rapidly progressing neurological deficits such as conus – or cauda – syndromes, or severe paresis (Grade II–0), segmental instability, pregnancy, litigation or psychogenic aggravation as well as all forms of non-discogenic radicular irritation.

Morphological contraindications are subligamentous prolapses which exceed 1/3 of the sagittal diameter of the spinal canal or which have migrated cranially or caudally to the disc space level. They include all forms of sequestrated disc herniations as well as all kinds of bony nerve root entrapment.

Table 2. *Contra-indications to Percutaneous Endoscopic LASER Discectomy*

Clinical aspect	Morphological aspect
Severe neurological deficits	Subligamentous prolapse
– Paresis (II°–0°)	– cranial/caudal to disc-space level
– Conus/cauda syndromes	– > 1/3 of sagittal diameter of spinal canal
Segmental instability	Epidural prolapse
Pregnancy	Fragments
Litigation	(Narrow lateral recess)
Psychogenic aggravation	(Narrow spinal canal)
	(Spondylolisthesis)

Surgical Technique*

The technique of Percutaneous Endoscopic Discectomy (PED) has been described in detail elsewhere[9]. Since the use of LASER requires a bilateral approach, the major steps of the procedure are briefly recalled here:

Percutaneous Endoscopic LASER Discectomy (PELD) is performed under local anaesthesia. Although general anaesthesia is contra-indicated in PELD, there should always be an anaesthesiologist in "stand by" in order to control the vital parameters, to support, if necessary, analgesic treatment, and to control rare, but possible vascular complications. The patient is placed in a prone, comfortable position on a radiolucent table. The level of the disc space to be treated is marked on the skin under fluoroscopic control. The disc space is punctured with an 18-Gauge-cannula through a postero-lateral approach at an angle of 50–60°, entering the skin between 9 and 10 cm from the midline. The tip of the needle is placed into the center of the disc. Discography is performed to exclude morphological contra-indications such as sequestration. However, in our experience contra-indications could be excluded by pre-operative imaging techniques in about 90% of our patients. This means that in only 10% of the cases, the indication for PELD had to be revised due to the result of discography. The disc space is now punctured from the contralateral side in the same manner. Two guide wires are introduced into the disc space through the cannulas which are then removed. Two stab incisions are performed at the entry points of the wires. A blunt trocar with a central burr hole is then advanced to the posterolateral border of the disc space using the wire as a guide. The cone of the trocar is detached, and a working cannula (OD: 5.4 mm) is introduced using the trocar as a guide (Fig. 1). The disc space is then opened from each side with an annulus trephine. Parts of the nucleus pulposus are removed from the center of the disc so as to create a hollow space in which endoscopic inspection is possible. Removal of disc material can be accelerated by using automated shavers and cutters[9]. A 30 or 70° rigid endoscope is introduced into the disc space from one side, while removal of nucleus pulposus is continued under video-endoscopic control from the other (Figs. 2, 3 and 4).

---

* Instruments: Aesculap AG, Tuttlingen, Germany.

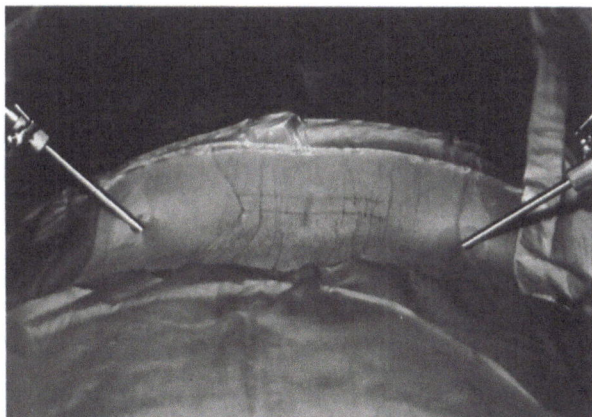

Fig. 1. Working cannulas introduced at L 4/5 through a bilateral postero-lateral approach

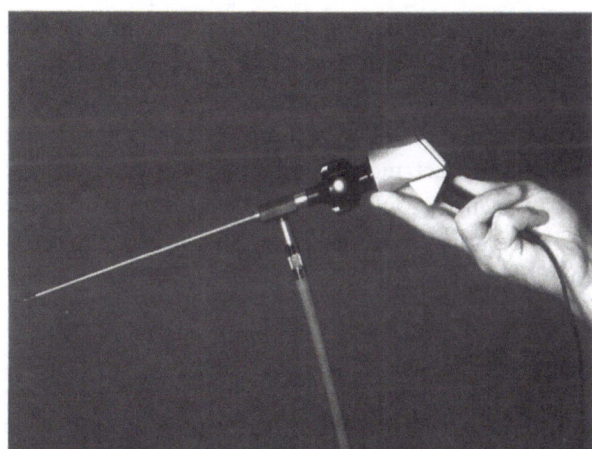

Fig. 2. Rigid endoscope connected to a chip-video-camera

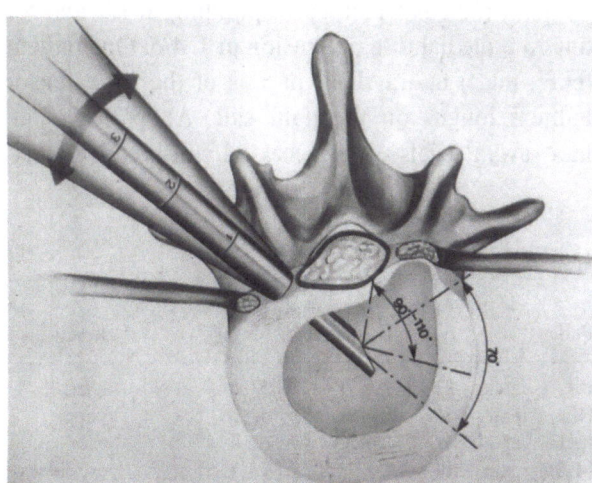

Fig. 3. Visual field of a 70° endoscope introduced into the disc space

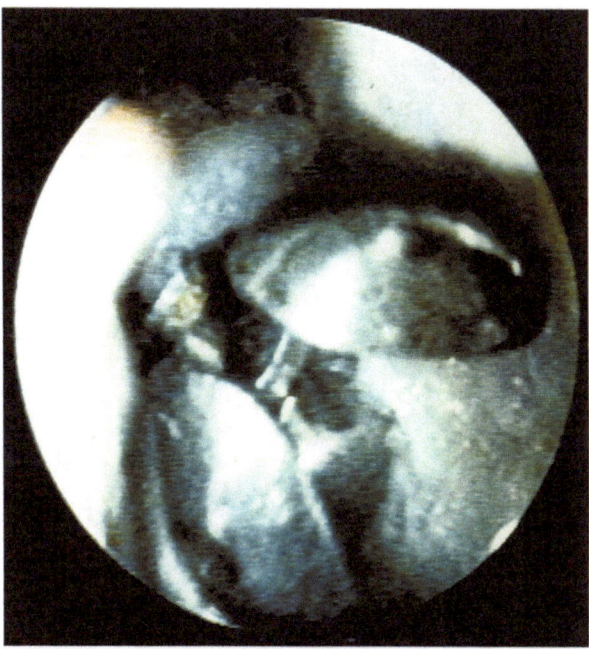

Fig. 4. Endoscopic view of the forceps introduced into the disc space

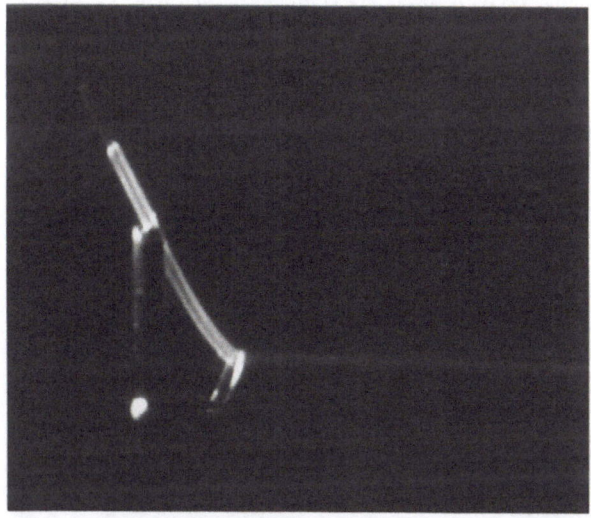

Fig. 5. 400 µm quarz-fiber with adjustable deflector

In order to reach the posterior third and the dorsal, subligamentous border of the intervertebral space, a flexible 400 µm quarz-fiber ("bare fiber") is introduced into the disc space using an adjustable deflector (Fig. 5). This fiber is coupled to a cw-1064 nm-Nd:YAG-LASER (Axyon GL 001, Aesculap-Meditec, Heroldsberg, Germany) (Fig. 6). Nucleus pulposus tissue is then coagulated or vaporized under endoscopic control (Fig. 7). Smoke from tissue vaporization is removed by continuous suction through the working tube. The patient is occasionally asked to press or cough. During this manoeuvre, loose nucleus

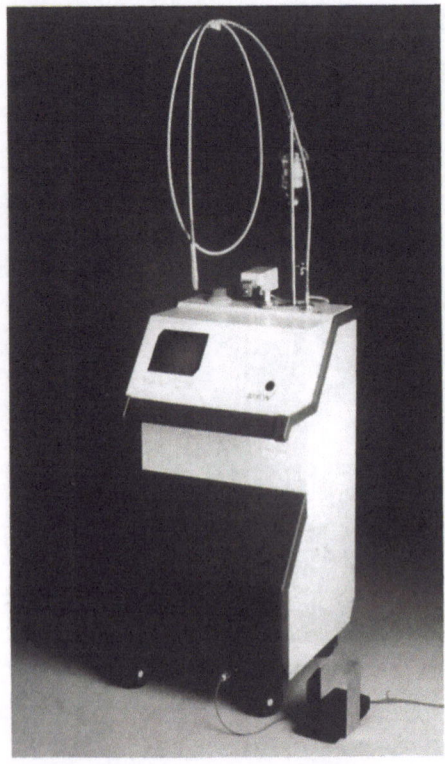

Fig. 6. Nd:YAG 1064 nm-LASER (Axyon GL001, Aesculap A, Tuttlingen, Germany)

Table 3. *LASER-Parameters Used for Percutaneous Endoscopic LASER Discectomy*

	cw-Nd: YAG-LASER 1064 nm
Application-power	20–30 Watts
Pulse–duration	0.05–0.1 sec
Bare fiber	0.4 mm
Contact mode	–

LASER-Parameters (Table 3)

The experimental basis for the endoscopic use of the Neodymium:YAG-LASER in the disc space will be described elsewhere[10]. The LASER can be used either in a non-contact or in a contact mode. The latter has been shown to be more effective using application powers between 20–30 Watts, and pulses between 0.05 and 0.1 sec. By using a 400 μm bare-fiber energy densities between 1.2 and 2 KW/cm2 could be applied in a non-contact mode with a tissue-fiber-distance of 2 mm. However, only coagulation and shrinking of the nucleus pulposus can be achieved with the non-contact mode. By using the contact mode, energy densities of between 10 and 15.6 KW/cm^2 could be applied to the tissue. This resulted in effective vaporization.

Preliminary Results (Table 4)

Between August 25th, 1989 and April 4th, 1990 6 patients (4 male, 2 female) were treated by PELD. All had been candidates for Percutaneous Endoscopic Discectomy by the criteria mentioned initially. All had radicular pain and sensory disturbances corresponding to the nerve roots L 5 and/or S 1. The two women had bilateral radicular pain corresponding to L 5 and S 1 due to a medial disc protrusion at L 4/5. One patient (H.F., male) had a slight paresis of the M. extensor hallucis longus on the right side. All patients had had several trials of unsuccessful conservative treat-

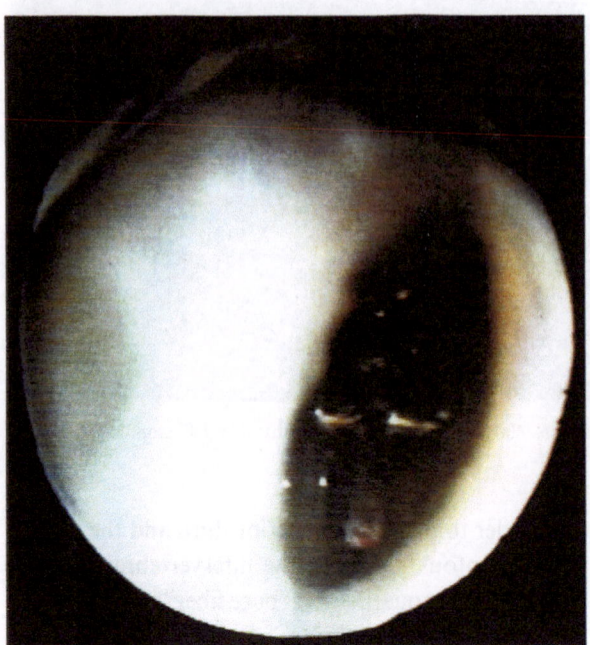

Fig. 7. Endoscopic view of the quarz-fiber introduced into the intervertebral space

pulposus tissue can be identified in the dorsal part of the disc space and selectively removed with the LASER. The procedure is completed when no more loose nucleus pulposus tissue can be identified at the posterior margin of the intervertebral space.

Table 4. *Patients Data*

Patient		Age (y)	Disc-level	Hospitalization (days)	Result
P.J.,	male	23	L 4/5 l	5	good
H.F.,	male	55	L 4/5 r	5	good
R.H.,	male	41	L 4/5 l	6	excellent
M.A.,	male	19	L 4/5 r	2	excellent
D.R.,	female	27	L 4/5 med	5	good
W.S.,	female	30	L 4/5 med	10	satisfactory

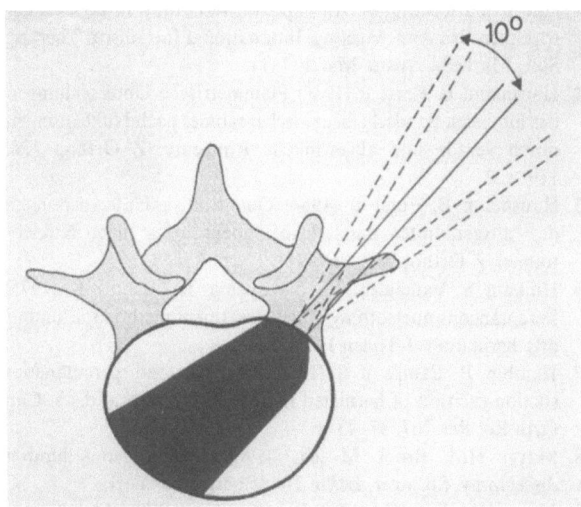

Fig. 8. Working range of rigid forceps introduced through a postero-lateral approach

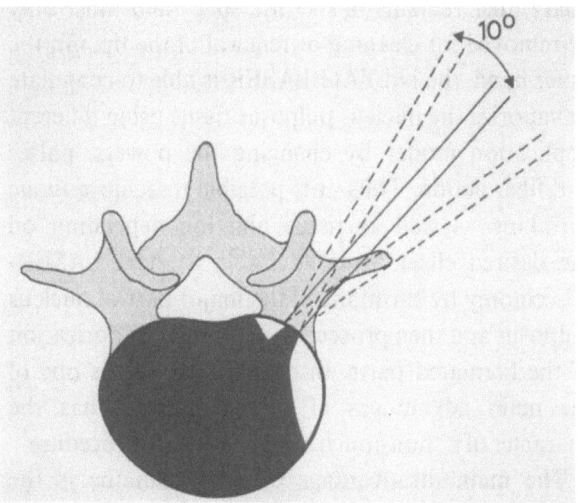

Fig. 9. Working range of a flexible LASER-quarz-fiber introduced through postero-lateral approach

ment in the hospital. PELD was performed using the method described above. There were no complications. All patients were free of symptoms or considerably improved the day following the procedure. Hospitalization was between 2 and 10 days. At the day of dismissal, all patients were free of radicular symptoms. Three patients complained of slight stress-dependent back-pain. There were no side effects attributable to LASER in the post-operative period.

Discussion

The goal of Percutaneous Discectomy is to reach any given point inside the disc space, and to remove the herniated portions of the nucleus pulposus in the dorsal part of the intervertebral space. The intradiscal working-range of rigid forceps used for microdiscectomy or for percutaneous endoscopic discectomy permit reaching only about 60% of the intervertebral space[4,5] (Fig. 8). By entering the disc space through a postero-lateral approach from one side, this working range can be extended to 93% using the flexible quarz-fiber for LASER application (Fig. 9). Since the instruments need room to be moved, it is necessary to remove central parts of the nucleus pulposus before introducing the deflector and the quarz-fiber. This can be achieved with the manual and automated instruments for Percutaneous Endoscopic Discectomy[7,8,12]. There were no difficulties in moving and directing the quarz-fiber inside the disc-space towards the nucleus-pulposus targets. There were no complications due to the instruments although a fiber breaking was observed in one patient. However, since

the procedure is performed under endoscopic control, the fiber fragment could be removed without problems using the small rongeur. Although our clinical experience with the additional use of LASER during Percutaneous Endoscopic Discectomy is limited, we consider PELD to be safe since the use of the instruments is controlled by endoscopy and the safety range of the LASER parameters has been defined according to our experimental results[10].

In 1988 ASCHER described the ND:YAG-LASER denaturation of nucleus pulposus using a needle technique[1,2]. The quarz-fiber is introduced into the disc space through a cannula. Nucleus pulposus is coagulated *without endoscopic control*. Since the fiber is not steerable, only the central part of the nucleus pulposus can be coagulated, thus achieving an "intradiscal decompression". This local effect is comparable to the so-called "Automated Percutaneous Lumbar Discectomy (APLD) – method using an oscillating knife in a rigid 2 mm – working-cannula ("Nucleotome") coupled to a suction-irrigation-system. These methods do not allow selective nucleotomy as has been shown experimentally[3,13].

In order to include small subligamentous prolapse into the spectrum of indication for Percutaneous Discectomy, the use of flexible instruments as well as of video-endoscopic control is mandatory[9,12]. Whether the complementary use of LASER is superior to the use of flexible forceps is open to discussion. Our experiences are preliminary and do not allow final conclusions. However, the flexible forceps we use for PED are limited in size so that the removal of the dorsal portions of nucleus pulposus is sometimes time consuming and difficult. Using the LASER, the quarz fiber remains in the disc space and must only be removed for cleaning or renewal of the tip. On the other hand, the Nd:YAG-LASER is able to coagulate or vaporize the nucleus pulposus tissue using different application modes by changing the powers, pulses and fiber action. Thus, it is possible to achieve tissue shrinking as well as tissue ablation depending on the desired effect. It is advisable to start LASER-Discectomy by shrinking of the major part of nucleus pulposus and then proceed with selective vaporization of the herniated parts. In our opinion this is one of the main advantages of PELD, since it has the character of a "non-touch" and "one-way" procedure.

The main disadvantage of this technique is the necessity of a second approach to the disc space. However, the development of a LASER-applicator with an integrated endoscope and a suction-irrigation-device is in progress.

References

1. Ascher WP, Holzer P, Claici G, Choy DS, Jury H (1988a) Denaturation and vaporization of the nucleus pulposus of herniated intervertebral discs. Int. Symposium on Percutaneous Lumbar Discectomy, Berlin, August 12–13
2. Ascher WP (1988b) Laser in der Neurochirurgie. In: Walter W, Krenkel W (Hrsg) Jahrbuch der Neurochirurgie. Regensberg & Biermann, Münster, pp 101–119
3. Griss P (1990) Effects of automated percutaneous lumbar discectomy on intervertebral disc tissue compared with non-automated techniques: An experimental study in 60 cadaver specimen. 3rd Ann. Meeting, International Intradiscal Therapy Soc., Marbella, Spain, March 7–11
4. Hausmann B, Forst R (1986) Planimetrische Untersuchungen des lumbalen Bandscheibenzwischenraumes nach Nukleotomien durch gerade und abgewinkelte Rongeure. Z Orthop 124: 119–122
5. Hausmann B, Forst R (1984) Quantitative Untersuchungen des "ausgeräumten Bandscheibeninnenraumes" nach Nukleotomien. Z Orthop 122: 269–271
6. Hijikata S, Yamagishi M, Nakayama T, Oomori K (1975) Percutaneous nucleotomy: A new treatment method for lumbar disc herniation. J Toden Hosp 5: 39
7. Kambin P, Sampson S (1986) Posterolateral percutaneous suction-excision of herniated lumbar intervertebral discs. Clin Orth Rel Res 207: 37–43
8. Mayer HM, Brock M (eds) (1989) Percutaneous lumbar discectomy. Springer, Berlin Heidelberg New York
9. Mayer HM, Brock M (1989) Percutaneous lumbar discectomy – the Berlin technique. In: Mayer HM, Brock M (eds) Percutaneous lumbar discectomy. Springer, Berlin Heidelberg New York, pp 107–117
10. Mayer HM *et al* (1992) Experimental investigations on the use of Nd:YAG-LASER on lumbar disc tissue. (In preparation)
11. Onik G, Maroon JC, Davis GW (1989) Automated percutaneous discectomy at the L5-S1 level. Clin Orth Rel Res 238: 71–76
12. Schreiber A, Suezawa Y, Leu HJ (1989) Does percutaneous nucleotomy with discoscopy replace conventional discectomy. Clin Orth Rel Res 238: 35–42
13. Shepperd JAN (1988) Personal communication

Correspondence: Priv.-Doz. Dr. HM Mayer, Neurochirurgische Klinik Universitäts-Klinikum Steglitz, Freie Universität Berlin, Hindenburgdamm 30, D-1000 Berlin 45, Federal Republic of Germany.

Acta Neurochirurgica, Suppl. 54, 59–62 (1992)

Neuroendoscopic Technique for the Operative Treatment of Septated Syringomyelia

N. Huewel, A. Perneczky, V. Urban, and G. Fries

Neurosurgical Department of the Johannes Gutenberg-University, Mainz, Federal Republic of Germany

Summary

The management of septated, chambered syringomyelia has until now been problematic because the usual operative methods cannot secure drainage of all chambers of the cavity.

The development of a flexible neuroendoscope affords the possibility to perforate the septa under visual control which sub-divide the cavity.

We present our experiences with 11 cases of septated syringomyelia, which have been operated upon using a flexible neuroendoscope.

Keywords: Syringomyelia; syringo-subarachnoid shunt; septum perforation; flexible endoscope.

Introduction

The results of the various operational techniques in cases with polyseptated, chambered syringomyelia are still not satisfactory. A microsurgical myelotomy – over a limited laminectomy – does not always guarantee a free communication of all chambers.

In order to perforate the septa and secure this connection under optical vision and control with a minimal approach a microneuro – endoscopic technique was developed and established. This method has been applied in 11 cases with satisfactory results.

Material and Methods

Our surgical material consists of 11 cases. The first operation was performed on late 1989. So far we have operated on 7 cases with idiopathic syringomyelia, which were polyseptated and polychambered and had cervico-thoracic extension without hindbrain malformation, 3 posttraumatic syringomyelias after compression fractures of thoracic vertebral bodies and 1 case with associated syringomyelia due to an intramedullary astrocytoma Th 5–10.

First a laminectomy of 1 to $1\frac{1}{2}$ levels at the caudal end of the syringo- or hydromyelia is performed. Following incision of the dura over 5–7 cm, at the same time splitting of the arachnoid, is carried out. Then myelotomy will be performed in the fissura mediana dorsalis by microsurgical technique; an opening of about 2.5 mm in length is sufficient for the application of the neuroendoscope.

Our endoscopic equipment consists of 3 fiber endoscopes which are at present in the development phase for neurosurgical requirements:

Fiberendoscope	– with an outside diameter of 0.85 mm (2.5 F), 100 cm length, with camera-capable ocular and connection for cold light cable, 6000 image points.
Fiberendoscope	– as above, but with connected working channel for irrigation, suction, and uptake of the laser fiber.
Fiberendoscope	– 2.3 mm outside diameter (7 F), 80 cm length, with integrated 1 mm working channel, camera-capable ocular and connection for cold light cable.
CCD-PAL-camera	– with external camera control unit and 38 mm objective.
Cold light source	– 300 Watt, xenon with cold light cable.
PAL-monitor	– with high resolution monitor.

The flexible fiber endoscopes are gas-sterilised in ethylene dioxide 50 °C, 12 h exposure time, 8 h evaporation time. Dependent on the target, diameter of syrinx or the distance from approach to the operation-field, either the 0.85 mm or the 2.3 mm neuroendoscope is used.

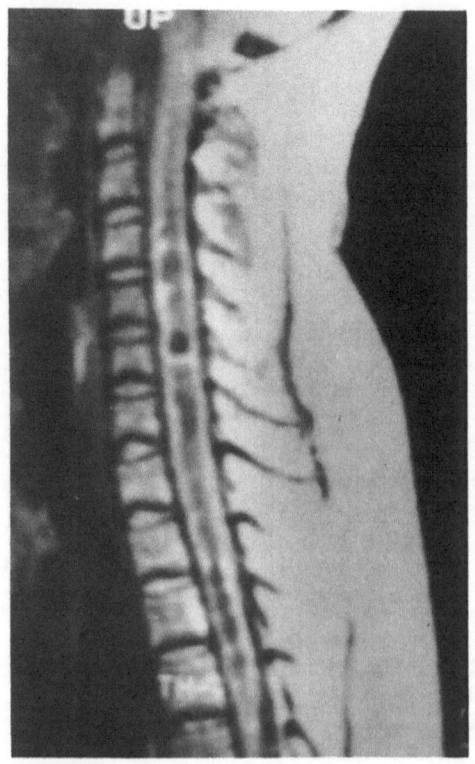 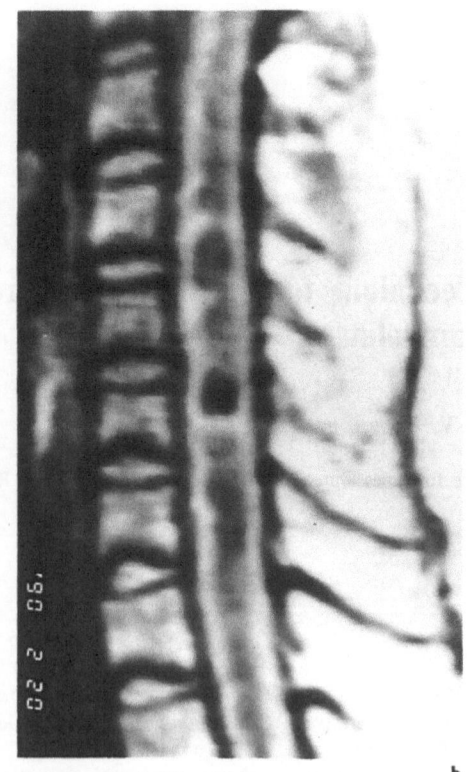

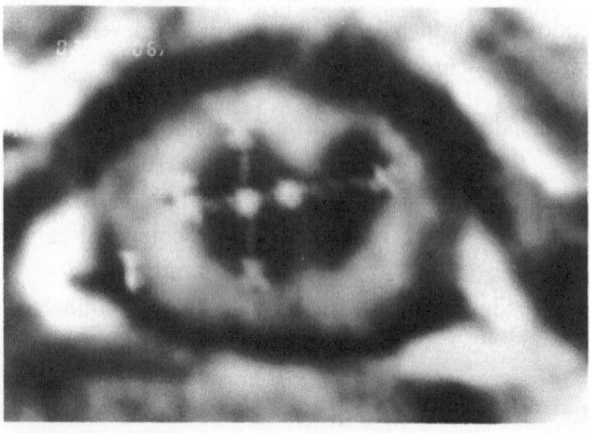

Fig. 1 a–c. Multiple and partly solid chambered cervico-thoracic syringomyelia in a 38 year old female patient, pre-operative MR findings

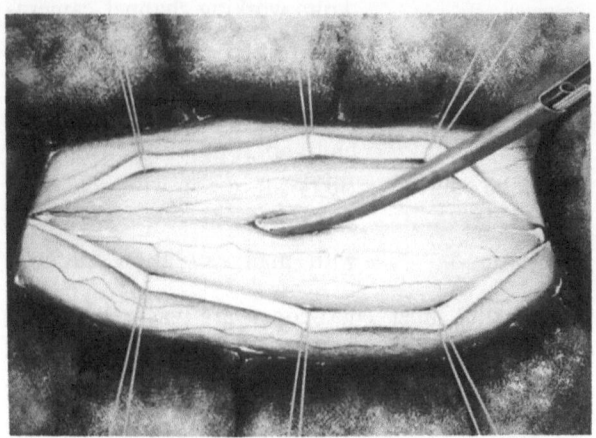

Fig. 2. Introduction of the 0.85 mm microendoscope with connected working channel of also 0.85 mm diameter

Under demonstration on the high resolution monitor, the septa are reached and, if necessary, perforated, at poorly vascularised regions which are chosen for perforation. The septa can be communicated either by mechanical or laser-assisted methods. Intrasyringeal vessels which are injured can be dealt with by irrigation, adhesion or with the laser. For completion of the neuroendoscopic equipment we are engaged in the development of a holmium-laser adapted to neurosurgical problems, which can be transmitted over shoulder- or gradient fibers with a diameter up to 0.1 mm. After control of the complete intracavernal syringostomy a silicon continuous drainage is led from the syrinx to the subarachnoid

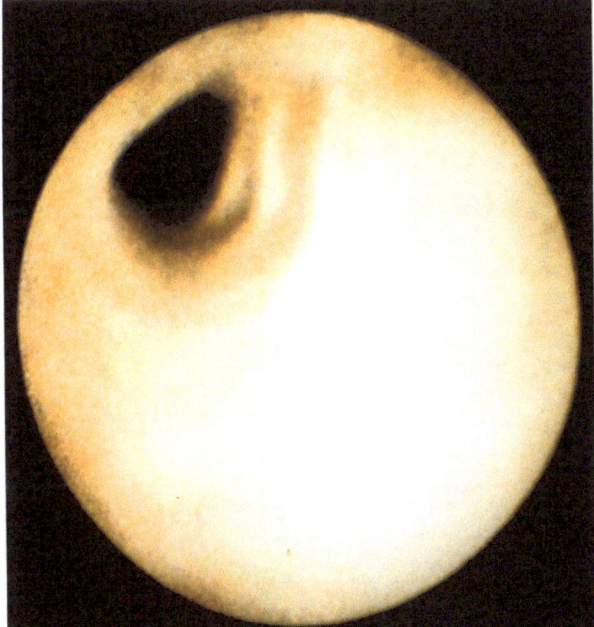

Fig. 3. View in an unseptated part of the syrinx

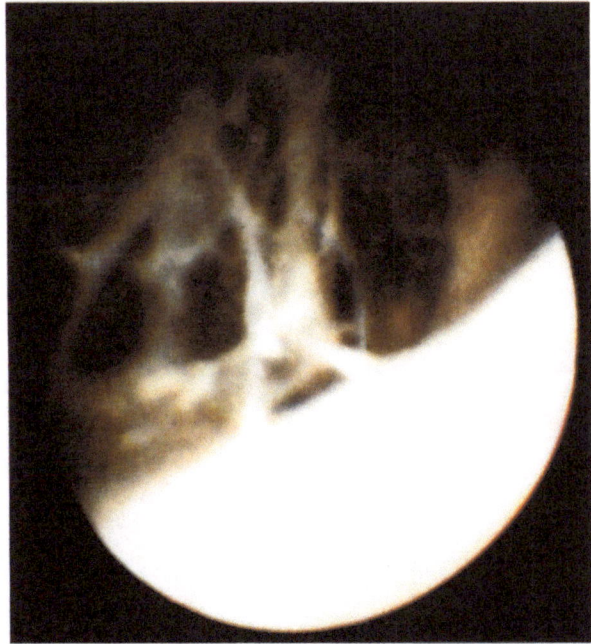

Fig. 4. Filigree fibres and small vessels

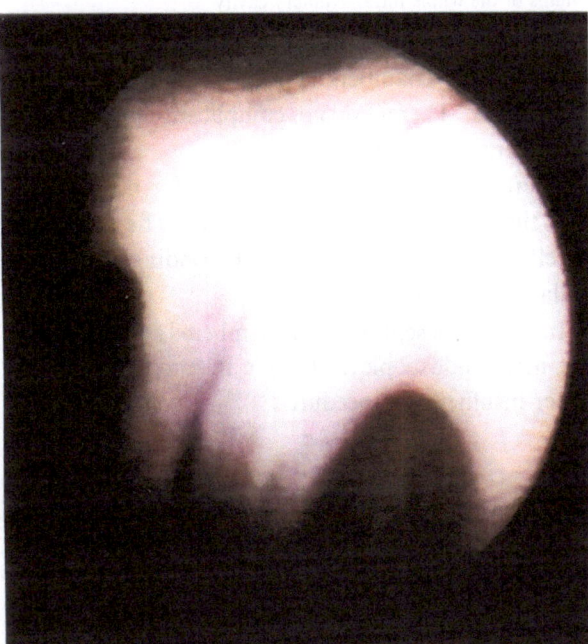

Fig. 5 a. Solid middle bar septum with 3 lumina; vascular distribution within the syrinx

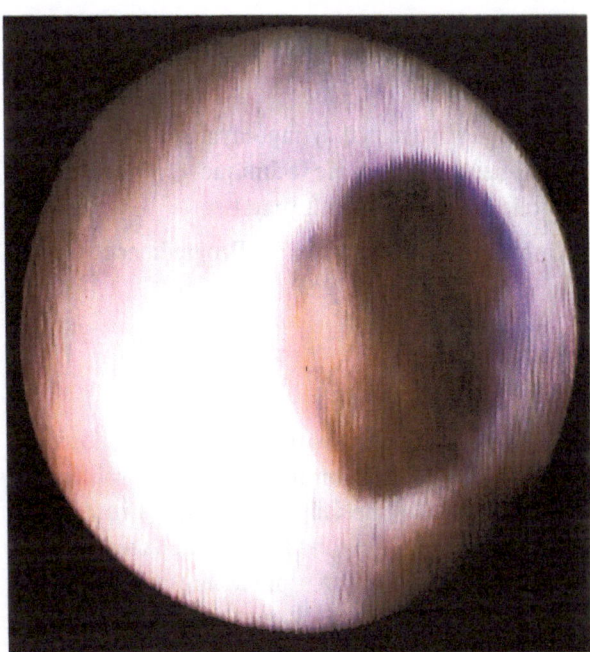

Fig. 5 b. Perforated septum

space in the usual way and is then fixed. The additional operation time of extended, often septated syringomyelias is about 60 min. The neuroendoscopy of syringomyelia is indicated even when other pre- and intra-operative methods do not clearly exclude a tumorous concomitant syringomyelia.

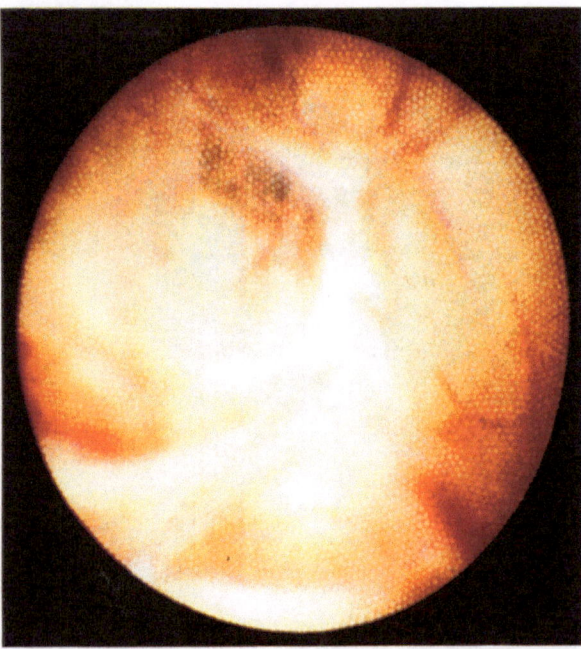

Fig. 6. Endoscopic view of the region of the obex in a case of communicating syringomyelia. Endoscope 38 cm above a myelotomy at Th 10 pushed cranial wards

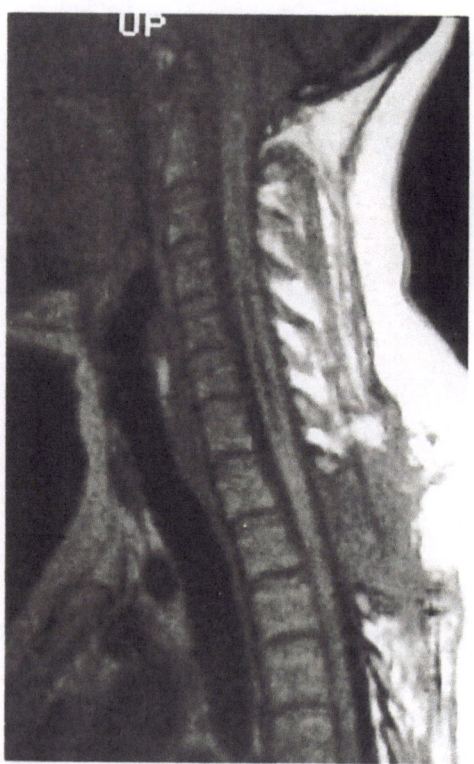

Fig. 7. Check NMR two days postoperatively demonstrates a collaps of the whole intramedullary cavity

Results

11 surgically treated syringomyelias
with neuro-endoscopic technique
(Follow up: 1 month–2 years)

Improvement	Stable	Progredience
9	2	0

Discussion

The neuroendoscopy with flexible fiber endoscopes has proved its indication in therapeutic problems in the spinal canal. Especially syringomyelia in its poly-chambered form is an indication for using the new technique. Perforating all septa secures the communication of all chambers. The procedure is safe and atraumatic. There are no complications intra- and postoperatively.

There is no alternative procedure for making sure that the cavity is completely drained. The results are excellent, the postoperative MRI-findings show the collapsed syrinx in all cases.

In cases with septated and chambered syringomyelias the neuroendoscopic technique is to be preferred to other insufficient procedures.

Correspondence: Dr. N. Huewel, Neurosurgical Department of the Johannes Gutenberg-University, D-W-6500 Mainz, Federal Republic of Germany.

Acta Neurochirurgica, Suppl. 54, 63–68 (1992)

Minimally Invasive Neurosurgery by Means of Ultrathin Endoscopes

D. Hellwig and **B.L. Bauer**

Department of Neurosurgery, Philipps-University Marburg, Federal Republic of Germany

Summary

The term "minimally invasive neurosurgery" (MIN) is defined and the present indications for MIN are described. They include endoscopic stereotactic interventions, endoscopic evacuations of chronic subdural haematomas and intracerebral mass-haemorrhages, endoscopic spinal and ventriculoscopic procedures. The advantages compared to conventional diagnostic and therapeutic neurosurgical approaches consist of less operative risk and reduced tissue-traumatization. Endoscopy makes interventions under (real-time) conditions possible. The operative stress for the patient is minimized.

Keywords: Minimal invasive neurosurgery; endoscopy; stereotactic neurosurgery; chronic subdural haematoma; massive intracerebral haemorrhage; ventriculoscopy.

Introduction

The term "minimally invasive surgery" refers to interventions in which by application of modern technologies, large surgical opening of body cavities can be avoided[4,27].

With the help of endoscopes and newly developed miniaturized instruments complex surgical interventions can be carried out. In neurosurgery there is a similar tendency to less invasive surgery. In analogy to the term "minimally invasive surgery" we call these procedures "minimally invasive neurosurgery" (MIN). The present indications are endoscopic stereotactic interventions, evacuations of chronical subdural haematomas and intracerebral haemorrhages, endoscope-guided intraspinal procedures and ventriculoscopic interventions.

Material and Methods

1. Instruments (Figs. 1 a–d)

The main problem with the use of endoscopes in brain sugery is the size of the instruments. Their relatively large diameters can lead to considerable brain trauma. Another difficulty is how to use a flexible instrument in solid brain tissue. For these reasons most neuroscopes or encephaloscopes are rigid. For intracerebral and intraspinal interventions, however, it would be better or less traumatic, to use flexible endoscopes.

1.1. Flexible steerable endoscope: The *endoscope* we use is a prototype which was developed by Olympus Inc. Tokyo based on our experience with endoscopic neurosurgery. The specifications of the endoscope are listed in Table 1. Biopsy forceps and microscissors are adapted to the endoscope. Further micro-instruments are under development.

1.2. Endoscope-guiding-system: We have developed a special *endoscope-guiding system* to apply flexible, steerable endoscopes to problem areas of interest in conventional and stereotactic neurosurgery. The guiding-tubes consist of teflon. This material is relatively rigid and transparent in order to offer a direct view on the way to the target area which will be operated on. The tubes are adapted to the Mundinger-Riechert stereotactic system and to the self-retractor arm.

1.3. Intra-operative video-recording and display: The operative procedures are transmitted permanently by an ultralight microchip-camera to a video-unit. In this way the surgeon has a direct (real time) pictorial-recording and a documentation of the intervention.

2. Operation-Technique

2.1. Endoscopic stereotaxy: CT-guided stereotactic procedures are done with the Mundinger-Riechert stereotactic frame. The operative-technique is described elsewhere[16]. After the calculation of the stereotactic parameters and the burrhole trephination, we insert the endoscope guiding-system. The endoscope is guided afterwards to the calculated target-area and the region is inspected. The biopsy-forceps is introduced and tissue specimens are taken. The histopathologist examines the biopsies using smear-preparation techniques. Cysts are emptied under stereotactic conditions. The endoscopic inspection of the cavities guarantees the direct intraoperative control of complete evacuation.

2.2. Drainage of chronic subdural haematomas: The operative technique of choice is burr-hole trephination with subsequent introduction of a subdural catheter. To achieve a complete evacuation of encapsulated chronic subdural haematomas, we perform endoscopic inspection of the subdural space in all directions and cut the membranes with microscissors to guarantee a free efflux of the haematoma-fluid.

Table 1. *Specification XURF-P2-40 (A7955)*

Optical System	Field of view	90°
	Direction of view	forward viewing
	Depth of field	1-50mm
Distal End	Outer diameter	3.1mm
Bending Section	Range of tip bending	180°/100° (UP/DOWN)
Insertion Tube	Outer diameter	3.3mm
Length	Working length	400mm
	Total length	710mm
Channel	Inner diameter	1.2mm

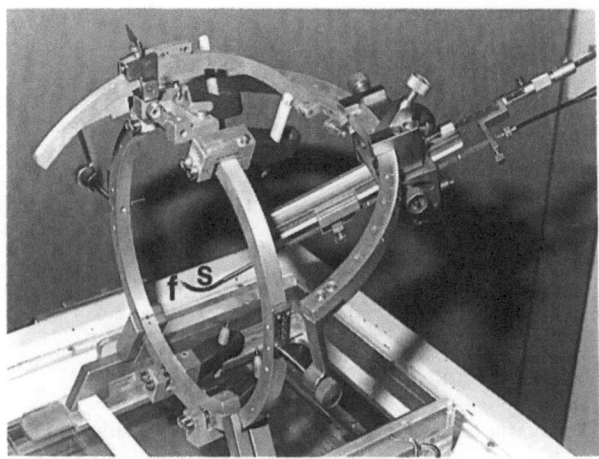

Fig. 1 c. Adaption of the endoscope to the stereotactic frame; biopsy forceps (*f*) is inserted in the steerable, flexible tip (*s*)

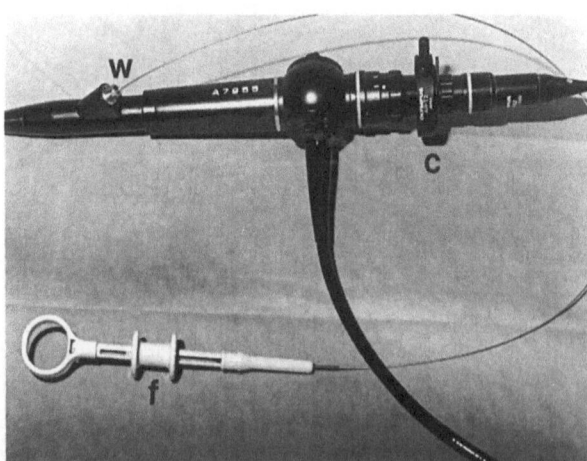

Fig. 1 a. Endoscope working-channel (*w*) with introduced biopsy-forceps (*f*) and connected camera (*c*)

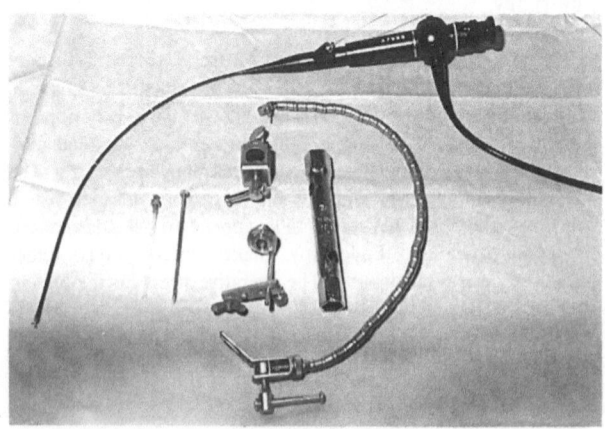

Fig. 1 d. Instrumentation for ventriculoscopy with the self-retractor arm

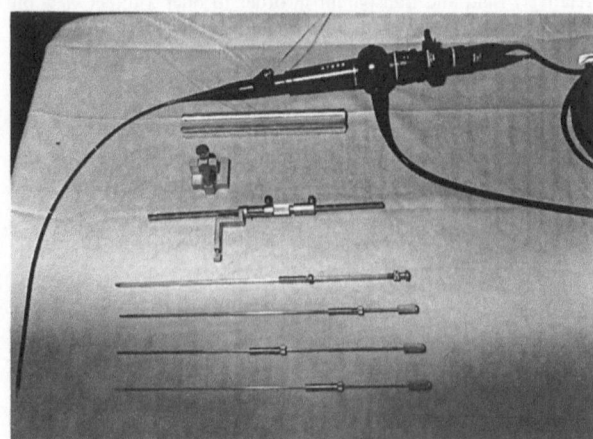

Fig. 1 b. Flexible, steerable endoscope with the equipment for stereotactic interventions

2.3. Intracerebral Haemorrhage: Intracerebral haematomas are operated on through a single burr-hole. First, the cavity is inspected endoscopically. The haematoma is evacuated under permanent rinsing and suction. Finally urokinase (Ukidan) 5000–10000 IE is injected into the remaining blood-clot and a silicon-drainage is installed. If the haemorrhage has ruptured into the ventricular system, a second ventricular drainage is performed in the ventricle opposite the bleeding.

2.4. Syringomyelia: The insertion of a syringo-subarachnoid shunt (SSS) with a Silastic catheter is an effective operative method in the treatment of syringomyelia without descent of the cerebellar tonsils[23,25]. The operative approach is done by a single-level laminectomy.

The operating microscope is used in the further course of the operation. Dura mater is incised and the dorsal median fissure is opened by laser. Through this artificial opening the flexible endoscope is introduced and the syrinx is inspected. Membranes in the cavity are cut with microscissors.

Finally the syrinx-fluid is drained by the insertion of a syringo-subarachnoid shunt. This endoscopic operation technique was developed and first described by Hüvel *et al.* The results are discussed in this issue[10].

2.5. Ventriculoscopy: We have developed a special device for ventriculoscopy, which is adapted to the self-retractor-system. From the frontal approach, the lateral ventricle is punctured with a teflon tube. The endoscope is introduced and the ventricular-system is inspected for pathological findings. If necessary, the biopsies are taken through the endoscope working-channel. The introduction of ventricular-drainage under endoscopic control is performed. In this way a direct intra-operative control of the correct positioning of the drainage is possible.

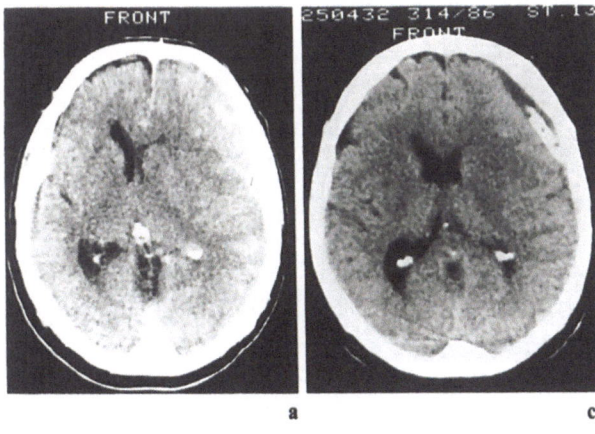

Fig. 2 a. CCT: Left-sided membraneous chronic subdural haematoma

Fig. 2 c. CCT-control after three days: Small residual haematoma after endoscopic evacuation with the applied external drainage

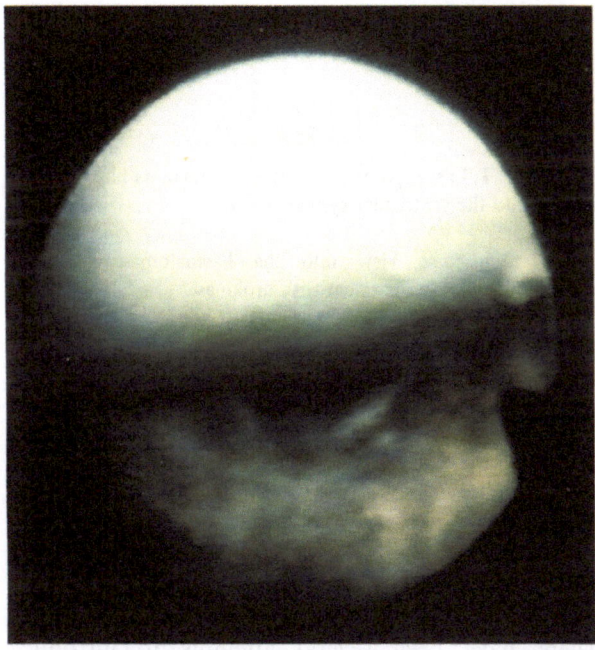

Fig. 2 b. Endoscopic visualization of the membranes inside the haematoma

Results

1. Endoscopic Stereotaxy

In 1989 we started using endoscopy during CT-guided stereotactic brain operations, mainly for biopsy procedures. So far we have operated on more than 50 patients with this technique.

The results of these interventions were generally positive. We can show the importance of having a direct view of the operative field, in order to recognize postbiopsy bleedings immediately and to carry out haemostasis under visual control. In cystic intracranial processes we have an intra-operative control about the totally emptied cyst without the necessity of further postoperative x-ray examinations. The mortality and morbidity of the interventions was zero. We have reported on this new operative technique elsewhere[8,9].

2. Drainage of Membraneous Chronic Subdural Haematoma

Successful treatment of chronic subdural haematomas can be achieved very easily. A simple burr-hole trephination and the institution of a subdural silicon drainage is the method of choice[13,22]. Problems occur if a subdural haematoma is divided in compartments by membranes. In those cases usually a large craniotomy is recommended. This procedure carries a high risk particularly because the chronic subdural haematoma usually occurs in elderly and weak patients.

We approach these therapeutic problems by endoscopic intervention. Through a small burr-hole trephination we enter the subdural haematoma. The membranes are cut by microscissors under visual control. In this way a drainage of the subdural haematoma as a whole is achieved (Figs. 2 a–c). The procedure is easy and without hazard for the patients. The results are satisfactory. Among 8 patients with membraneous subdural haematoma no re-operation was necessary.

3. Evacuation of Intracerebral Haemorrhage

Spontaneous intracerebral haemorrhages are a therapeutic problem in modern neurosurgery. In many cases the prognosis of hypertensive haemorrhages is poor even if a craniotomy with evacuation of the haematoma is performed. For this reason the treatment of choice is usually conservative or limited to neurosurgical techniques of a less invasive nature. We combined two minimal invasive approaches for our operative technique. First the endoscopic evacuation described by Auer[1] and secondly the intra-operative application of urokinase for lysis of coagulated intracerebral and intraventricular haematoma[15,24]. As yet we have not had much experience with this operative technique but the results with only a few patients are encouraging.

With the endoscopic-lytic treatment we are able to remove the main part of the haemtoma without craniotomy (Figs. 3 a–d).

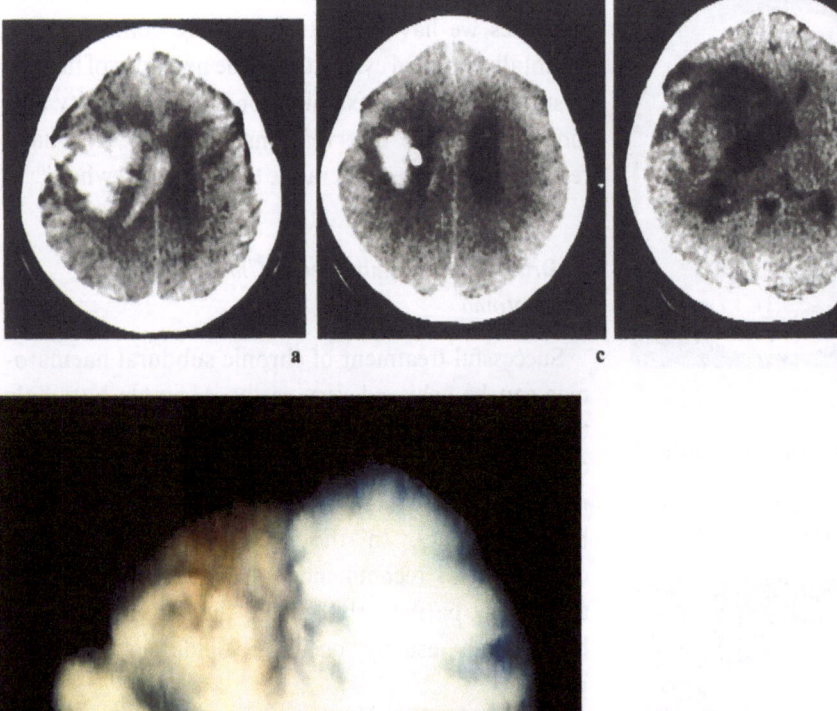

Fig. 3 a. CCT: Large spontaneous intracerebral mass-bleeding with rupture into the ventricular system

Fig. 3 b. Endoscopic view into the haematoma-cavity after evacuation and lytic treatment with urokinase

Fig. 3 c. CCT-control after one day: residual haematoma and external ventricular drainage

Fig. 3 d. CCT-control three weeks after the bleeding: The haematoma is evacuated. A large intracerebral defect is evident

4. Endoscopic Interventions on the Spine

A typical indication for the endoscopic interventions on the spine is the insertion of a syringosubarachnoid shunt (SSS) under visual control for treatment of syringomyelia. We have operated on only one case with this operative technique. In accordance with Huewel[10] the result was satisfactory (Figs. 4 a–d).

5. Ventriculoscopy

At the moment we only perform ventriculoscopy for inspection and biopsy of ventricle-related intracerebral processes. The main goal is to create an overview of the topographical relationships with respect to intra- and periventricular anatomical structures. An exact endoscopic mapping about the normal and pathological anatomy of the ventricular space is in preparation.

Discussion

There have been several attempts to introduce endoscopy in neurosurgery. In 1910 Dr. L'Espinasse from Chicago was the first to perform ventriculoscopic interventions for treatment of hydrocephalus[12]. This was followed by Dandy, Putnam and Scarf who carried out endoscopic interventions on hydrocephalus, partly by coagulation of the choroid plexus[5,20,21]. In 1921 Mixter succeeded in a third ventriculostomy under endoscopic control[14]. Third ventriculostomy using endoscopic operative techniques, was later successfully performed by Vries[26].

In 1986 Griffith summarized the activities in endoscopic intracranial surgery and called this operative technique *endoneurosurgery*[6]. The overall results of the endoscopic interventions at this time were moderate because of lack of suitable optical equipment and operative instrumentation. With

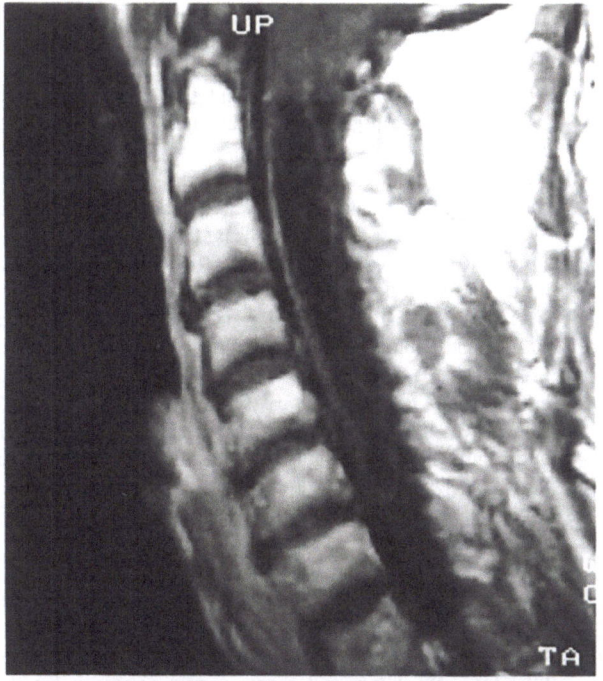

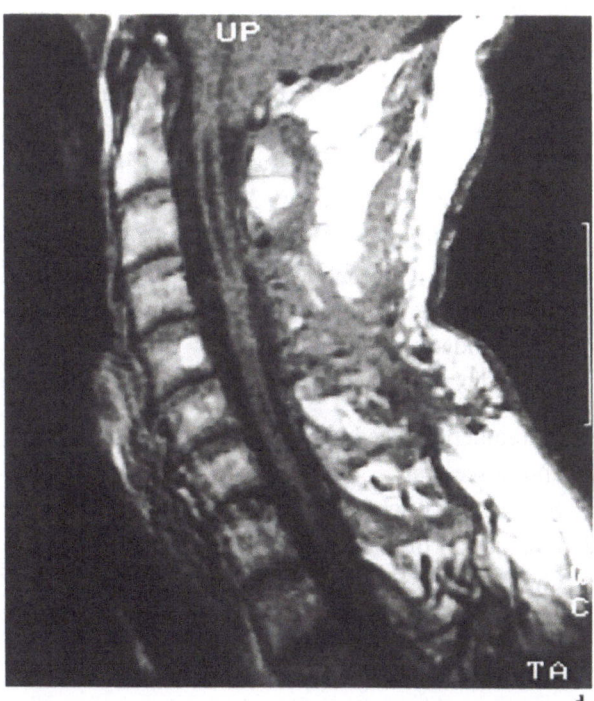

Fig. 4a. MRI: Syringomyelia of the cervical and upper thoracic spine

Fig. 4d. MRI-control after insertion of a syringo-subarachnoid shunt (SSS). Nearly the whole syrinx is drained. Two small cavities are left

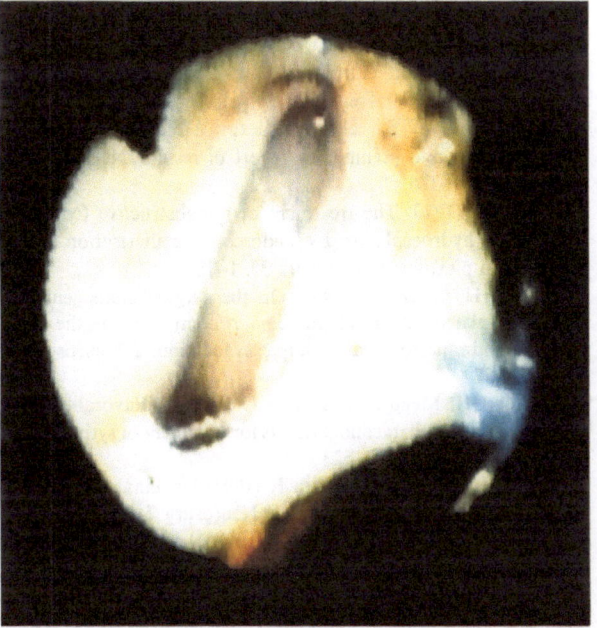

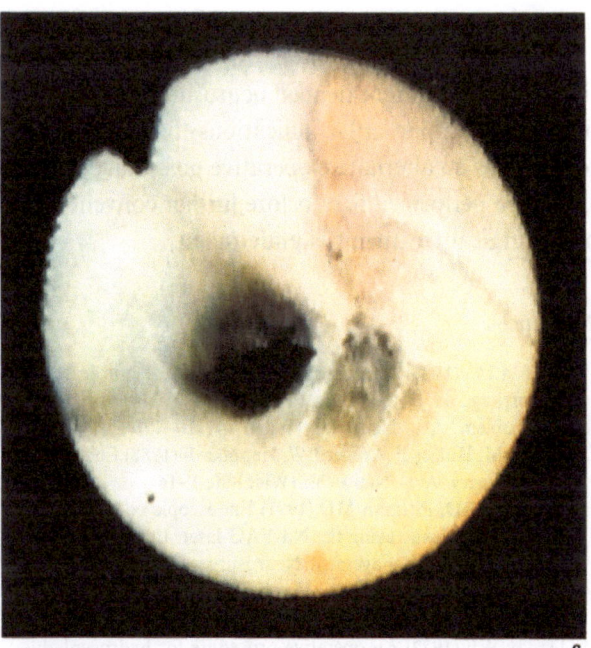

Fig. 4b. Artificial access to the syrinx through the dorsal median fissur

Fig. 4c. Membranes from inside the cavity

technical improvement of the optical systems and the development of endoscopic micro-instruments, endoscopy in neurosurgery has gained in importance and the indications widened. Today the main indications are the evacuation of intracerebral haematomas[1], diagnostic and therapeutic interventions on brain tumours[2,17] and cysts[18,19] endoscopic stereotaxy[8,9], applications for intraspinal processes[10] and again endoscopic procedures for treatment of hydrocephalus[3,7,11].

The results of these procedures are generally positive as the other papers in this issue confirm. Because the field of endoscopic neurosurgical interventions has widened we have introduced a new term to summarize all indications for these operative procedures. In similarity with the phrase *"minimally invasive surgery"* which is applied in general surgery we call it *"minimally invasive neurosurgery"(MIN)*. Minimally invasive neurosurgery covers interventions in which, by application of endoscopic operative techniques, a craniotomy or a larger surgical opening of the spine can be avoided.

Our indications for *MIN* are:

1. Stereotactic procedures (biopsy, evacuation of cysts and abscesses).
2. Evacuation of chronic subdural haematomas.
3. Evacuation of intracerebral haematomas.
4. Cystic processes or cavities of the spine.
5. Ventriculoscopic interventions.

The advantages for *MIN* in these particular indications are evident. Operative risk and tissue-trauma is minimized. There is little physical discomfort or distress for the patient and the quality of life is not reduced essentially after the intervention. The hospitalization-time for the patient is limited either.

Today, minimally invasive neurosurgical interventions do not exclude extended neurosurgical standard operations. In particular indications they have to be regarded as an alternative operative possibility, which should be performed first, before further conventional neurosurgical treatment is undertaken.

References

1. Auer LM (1985) Endoscopic evacuation of intracerebral hemorrhage. High-tec-surgical treatment - a new approach to the problem? Acta Neurochir (Wien) 74: 124–128
2. Auer LM, Holzer P, Ascher PW, Heppner F (1988) Endoscopic neurosurgery. Acta Neurochir (Wien) 90: 1–14
3. Buchholz RD, Pittman MD (1991) Endoscopic coagulation of the choroid plexus using the Nd-YAG-laser: Initial experience and proposal for management. Neurosurgery 28, No. 3: 421–427
4. Buess G (1990) Endoskopie - Von der Diagnostik bis zur neuen Chirurgie. Deutscher Ärzte-Verlag Köln, pp 18
5. Dandy WE (1922) An operative procedure for hydrocephalus. Bull Johns Hopkins Hosp 33: 189–190
6. Griffith HB (1986) Endoneurosurgery - endoscopic intracranial surgery In: Symon L *et al* (eds) Advances and technical standards in neurosurgery, Vol 14, No 5. Springer, Wien New York, pp 2–24
7. Heilman CB, Cohen AR, (1991) Endoscopic ventricular fenestration using a "saline torch". J Neurosurg 74: 224–229
8. Hellwig D, Eggers F, Bauer BL, Likoyiannis A (1990) Endoscopic stereotaxis: preliminary results. Stereotact Funct Neurosurg 54 + 55: 418
9. Hellwig D, Bauer BL (1991) Endoscopic procedures in stereotactic neurosurgery. Acta Neurochir (Wien) [Suppl] 52: 30–32
10. Huewel N, Perneczky A, Urban V, Fries G (1992) Neuroendoscopi Technique for the Operative Treatment of Septated Syringomyelia. Acta Neurochir (Wien) this volume, pp 59–63
11. Jones RFC, Stening WA, Brydon M (1990) Endoscopic third ventriculostomy. Neurosurgery 26: 86–92
12. L'Espinasse VL (1943) In: Davis neurological surgery, 2nd ed. Lea and Febinger, Philadelphia, pp 442
13. McKissock W, Richardson A, Bloom WH (1960) Subdural haematoma. A review of 389 cases. Lancet 1: 1365–1360
14. Mixter WJ (1923) Ventriculoscopy and puncture of the floor of the third ventricle. Preliminary report of a case. Boston Med Surg J 188: 277–278
15. Mohadjer M, Ruh E, Hiltl DM, Neumüller H, Mundinger F (1988) CT-stereotactic evacuation and fibrinolysis of hypertensive intracranial haematoma. Fibrinolysis 2: 43–48
16. Mundinger F, Birg W (1984) CT-stereotaxy in clinical routine. Neurosurg Rev 7: 219–224
17. Otzuki T, Jokura H, Yoshimoto T (1990) Stereotactic guiding tube for open-system endoscopy: a new approach for the stereotactic endoscopic resection of intra-axial brain tumours. Neurosurgery 27 No 2: 326–330
18. Powers KS (1986) Fenestration of intraventricular cysts using a flexible, steerable endoscope and the argon laser. Neurosurgery 18, No 5: 637–641
19. Powell MP, Torrens MJ, Horgan JG (1983) Isodense colloid cysts of the third ventricle: a diagnostic and therapeutic problem resolved by ventriculoscopy. Neurosurgery 13: 234–237
20. Putnam TJ (1934) Treatment of hydrocephalus by endoscopic coagulation of the choroid plexus: Description of a new instrument and preliminary report of results. N Engl J Med 210: 1373–1376
21. Scarf JE (1970) The treatment of nonobstructive (communicating) hydrocephalus by endoscopic cauterization of the choroid plexuses. J Neurosurg 33: 1–18
22. Svien HJ, Gelety JE (1964) On the surgical management of encapsulated subdural hematoma. A comparison of the results of membranectomy and simple evacuation. J Neurosurg 21: 172–177
23. Tator CH, Meguro K, Rowed DW (1982) Favorable results with syringosubarachnoid shunts for treatment of syringomyelia. J Neurosurg 56: 517–523
24. Todo T, Usui M, Takakura K (1991) Urokinase infusion for intraventricular haemorrhage. J Neurosurg 74: 81–86
25. Vaquero J, Martinez R, Salazar J (1987) Syringosubarachnoid shunt for treatment of syringomyelia. Acta Neurochir (Wien) 84: 105–109
26. Vries JK (1978) An endoscopic technique for third ventriculostomy. Surg Neurol 9: 165–168
27. Wickham J, Fitzpatrick J (1990) Minimally invasive surgery. Br J Surg 77: 721

Correspondence: Dr. D. Hellwig, Department of Neurosurgery, Philipps-University Marburg, Baldingerstrasse, D-W-3550 Marburg, Federal Republic of Germany.

Acta Neurochirurgica, Suppl. 54, 69–76 (1992)

Stereotactic Endoscopic Interventions in Cystic and Intraventricular Brain Lesions

L. Zamorano, C. Chavantes, M. Dujovny, G. Malik, and **J. Ausman**

Henry Ford Neurosurgical Institute, Department of Neurological Surgery, Detroit, MI U.S.A.

Summary

Image guided stereotaxis is an accurate and safe method of directing therapy to target volumes defined in two-dimensional (2D) multiplanes or three-dimensional (3D) perspectives using computer reconstruction of image data. The major limitations of stereotactic techniques are related to a lack of intraoperative visualization and direct monitoring of the procedures and to changes of intracranial coordinates after decompression of cystic lesions or aspiration of cerebrospinal fluid in the management of intraventricular lesions. Endoscopic laser stereotaxis (ELS) involves integration of rigid-flexible endoscopy and Nd-YAG laser to 3D-2D multiplanar image-guided stereotactic procedures. The major advantages of ELS include: direct intraoperative visualization, hemostasis, evacuation or resection assessment, and wide exploration of intracranial cavities or ventricles. The technique allows safe aspiration, biopsy, and resection or internal decompression of deep and subcortical intracranial lesions. ELS has proved to be safe and effective in the management of 76 clinical cases and appears to be a promising technique in the management of cystic and intraventricular lesions.

Keywords: Endoscopy; fiberoptic; stereotaxis; laser; imaging.

Introduction

Image guided stereotaxis has proven to be an accurate and safe method to approach intracranial lesions, allowing therapy to be directed at target volumes defined in two-dimensions and three-dimensions using computer reconstruction of image data[3,6,12]. With two-dimensional (2D) multiplanar or 3D imaging processing, the size, shape, and main axis of intracranial lesions can be determined along with the important surrounding anatomical structures[14]. Image preplanning goals include selection of the safest and optimal approach for diagnosis and therapy of intracranial lesions[15]. This information can be transposed into stereotactic space with mathematical accuracy. Adequate methodology with 1 mm accuracy can be achieved on both the X and Y axis and

1.5 mm in Z axis[6]. Nevertheless, major criticisms and limitations of image-guided stereotaxis have been the lack of direct visualization and intraoperative monitoring of the procedures, morbidity and mortality due to hemorrhages associated with intraoperative blindness, and changes in intracranial coordinates after aspiration of cystic cavities or intraventricular lesions. To overcome some of these limitations in addition to increasing the scope of clinical usefulness of image-guidance to a wider range of intracranial pathology, we have adapted flexible and rigid endoscopy and Nd-Yag laser to stereotaxis for management of cystic and intraventricular lesions.

Material and Method

Endoscopic Laser Stereotaxis (E.L.S.) Instrumentation

The basic instruments used are described in Table 1. These include endoscopic instrumentation, (rigid-flexible), stereotactic frames (Zamorano–Dujovny, Riechert–Mundinger, Fischer, Freiburg, West Germany) and a cannula adapter for stereotactic arc. Rigid neuroendoscopy is performed by 6 mm cannula and a zero degree rigid optics (Karl Storz, Culver City, CA); additional available optics of 30 and 90 degrees may be useful in some cases. Endoscopic rigid instrumentation includes different biopsy forceps (punch cup, alligator, etc.), hook scissors, and bipolar coagulating electrodes. Fiberoptic neuroendoscopy is performed with a flexible Olympus fiberendoscope (ENT-1T10) (Olympus Corporation, Strongsville, OH) and flexible instrumentation similar to the rigid endoscope. A videocamera and a beamsplitter allow simultaneous display on the optics and television monitor. A 1.8 mm Nd-Yag laser disposable fiber (SLT, Malvern, PA, USA) is introduced through the 2 mm channels of both rigid and flexible endoscopes. The Nd-Yag laser (Cooper 1800) can be used in conventional noncontact technique or with contact laser probes.

Methodology

Endoscopic laser stereotaxis is a three-step process: 1) image data acquisition, 2) 3D-2D image processing for stereotactic planning, and 3) intraoperative procedure.

Table 1. *Endoscopic-ND-YAG Laser Stereotaxis (ELS) Requirements*

Image data	CT, MRI, DA
Surgical planning	axial views
	2D multiplanar
	(CT, MRI console)
	3D-2D software (SNPS*)
Stereotactic system	Riechert–Mundinger (RM)
	Zamorano–Dujovny (ZD)
Cannula adapter to arc	
Cannula with 2 stop-cock and mandril	Storz (26162)
Rigid endoscope	Storz (26162A)
Flexible endoscope	Olympus ENT-1T10
Xenon light source (300 w)	Storz, Olympus
Nd-YAG laser	Cooper 1800
Video camera	Storz Mini 9000 CD
Laser fiber (600 micron)	SLT
Rigid and flexible instrumentation	Olympus, Storz

* SNPS: Stereotactic Neurosurgical Planning System.

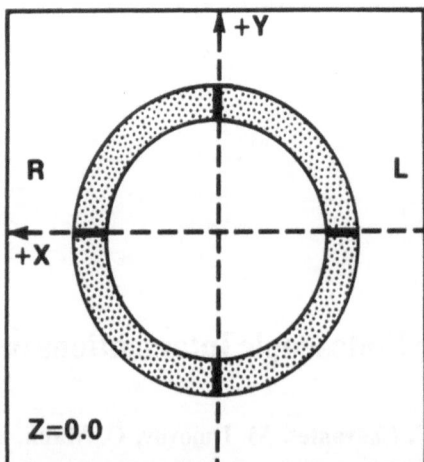

Fig. 2. Definition of the stereotactic coordinate system. The origin is the center of the ring plane

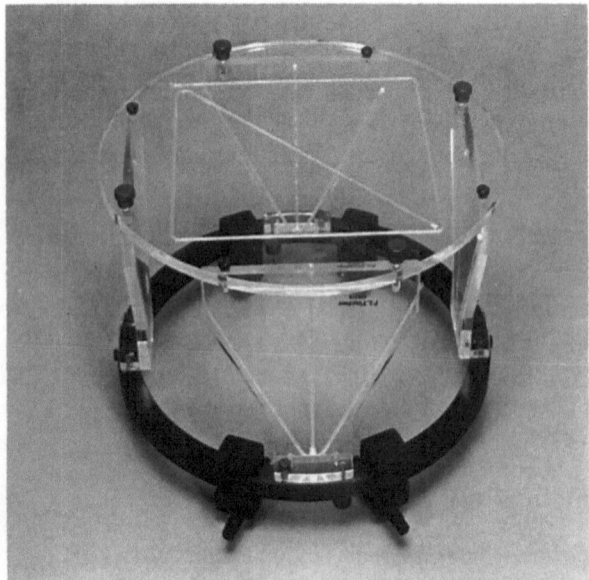

Fig. 1. ZD (Zamorano-Dujovny) base ring made of carbon fiber to allow CT. MRI and angiographic compatibilities. The attached localizer provides landmarks for CT and MRI

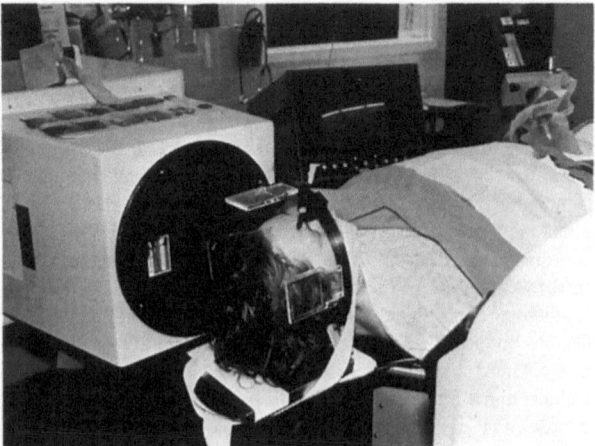

Fig. 3. Patient with ZD ring and angiographic localizers undergoing digital angiography

Image Data Acquisition

A computed tomography (CT) and magnetic resonance imaging (MRI)-compatible ring (ZD base ring, F.L. Fischer, Freiburg, West Germany) is fixed to the patient's head in the standard low position for supratentorial, diencephalic, and mesencephalic lesions. An inverted high position is used for lesions located in the infratentorial region or near the skull base. The fixation to the patient's skull is done by means of 3 or 4 pins that can be placed on any position to allow unobstructed intraoperative surgical approach (Fig. 1). The ring defines the stereotactic coordinate system; the origin is the center of the ring and coordinates are defined as x (right-left), y (anterior-posterior) and z (superior-inferior) to this origin (Fig. 2). The image data acquisition is performed with the fixed ring.

To interface the image data with the reference system, different approaches can be used depending upon the image modality, i.e. tomographic vs projective. In the case of tomographic imaging (CT or MRI) there are two main approaches: one alternative is the use of localizers (Fig. 1) attached to the ring which provide landmarks for stereotactic localization. The other alternative is the use of an adapter to the CT table or MRI coiler to bring the ring isocentric to the CT gantry or MRI coiler. In projective images (X-ray, angiography), thin fidutial radiopaque markers contained in four transparent plates are fixed to the ring (Fig. 3); a special alogorithm allows coordinates measurements considering the center of the ring as absolute origin independently of source-film distance or angulation.

3D-2D Image Processing for Stereotactic Planning

Basic 2D surgical preplanning can be performed at the CT console of the General Electric 9800 or at the MRI console of General Electric Sigma 1.5 with available software. Two-dimensional multiplanar preplanning can be performed with axial and reformatted views in coronal, sagittal, paraaxial, or oblique planes on CT images and with axial, coronal, and sagirral planes on cases of MRI. The location, size, shape, main axis, and anatomical and vascular relationships of the lesion can be visualized.

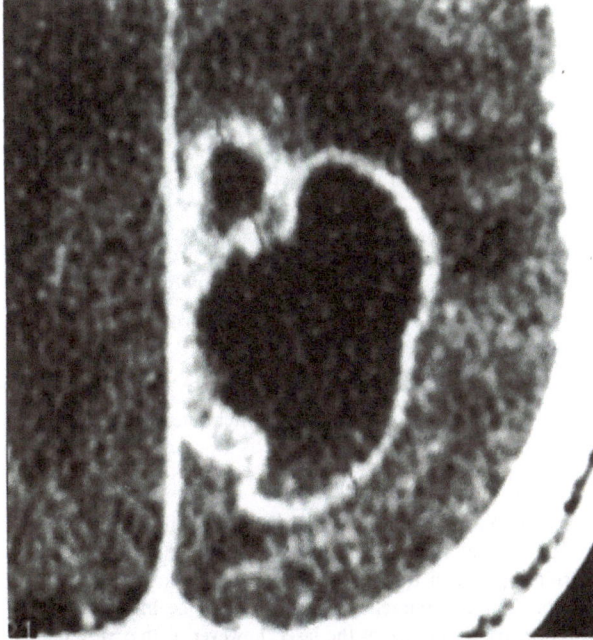

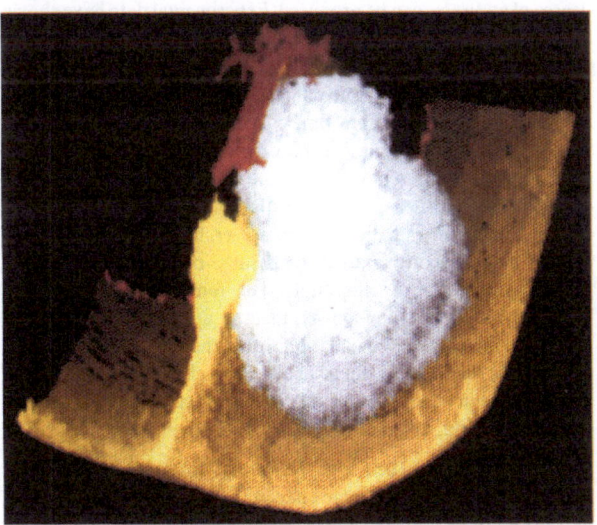

Fig. 4. A) Axial view showing biloculated cystic brain lesion; a trajectory has been selected to allow internal decompression of both compartments. B) 3D display of the same lesion under stereotactic conditions showing the margins and surrounding anatomical structures (vessels, dura, bone)

Ideally, stereotactic procedures involve multidimensional (3D and 2D multiplanar) interactive surgical preplanning. A "real-time" image based surgical planning utilizing multimodality imaging (CT, MRI, DA) has been developed in our Institution utilizing a Sun 4/370 workstation and specially developed hardware and software that provide 2D multiplanar and 3D views of the lesion for stereotactic procedures (SNPS, Stereotactic Neurosurgical Planning System, Henry Ford Neurosurgical Institute)[14,15]. The 2D multiplanar view includes reformatted vectors at any arbitrary plane of the patient's lesion such as sagittal, coronal, transverse, paraaxial, and free-tilt views, including oblique vectors. The 3D software includes features for extracting a view of the target volume localized by a process including steps of automatic segmentation, thresholding, manual contouring and boundry detection. Both 2D and

3D menus include "location" functions with "real-time" coordinate measurements and surgical trajectory definition capabilities as well as statistical functions for measuring distances, angles, areas and volumes. A combined interactive 3D-2D menu allows simultaneous display of selected trajectory, final optimization, and multiformat 2D display of free-tilt reformatted images perpendicular to selected trajectory of the entire target volume corresponding to the surgeon's eye view perspective (Fig. 4A and B).

For ELS preplanning, an automatic thresholding or manual contouring technique is used to delineate treatment areas on each of the axial slices. Treatment volume is calculated and displayed form any desired perpective and 2D multiplanar reformatted images are generated on coronal, sagittal, paraaxial, or oblique planes. The surgical trajectory of the rigid cannula of the endoscope is selected in 3D views, taking into consideration the tumour's axis, shape, size, number of loculations, vascularity, and surrounding anatomy. Generally, the surgical trajectory will follow the tumour's main axis unless anatomical or vascular considerations contraindicate such an approach. With the interactive 3D and 2D menu optimization and "simulation" of the procedure is performed. Finally, a batch of reformatted images is generated at any desired spacing intervel that corresponds to the planes prependicular to the endoscope's trajectory. The aim is to define the lesion and its anatomical relationships at different depths of introduction of the endoscope. The purpose is to select the best trajectory for the rigid cannula. In summary, two-point coordinates that define the optimized trajectory are generated at the time of preplanning and will be transposed into stereotactic space by using algorithm that calculates angles of the aiming device.

Intraoperative Surgical Procedure

Aiming Device Settings

In the case of the ZD aiming device (Zamorano–Dujovny multipurpose neurosurgical unit, F.L. Fischer, Freiburg, West Germany) the semiarc device can be mounted on any four quadrants of the base ring (anterior, right, posterior or left) according surgeon's preference allowing an unobstructed surgical approach[13]. The system is arc centered and the target x, y, z coordinates can be set directly on the device (Fig. 5). With special alogrithm an specific trajectory can be defined by calculating two angles at the specific possible mountings[10]. The probe holder on the aiming device adjust the instrument depth referred to mm above or below the target. A holder adapter for instruments between 4 and 7 mm is the one used for the endoscopic cannula.

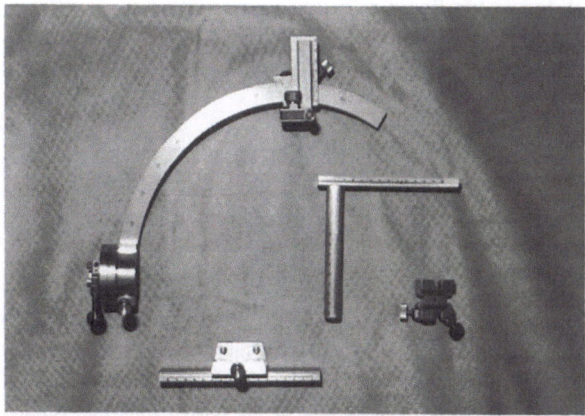

Fig. 5. ZD multipurpose localizing unit aiming device

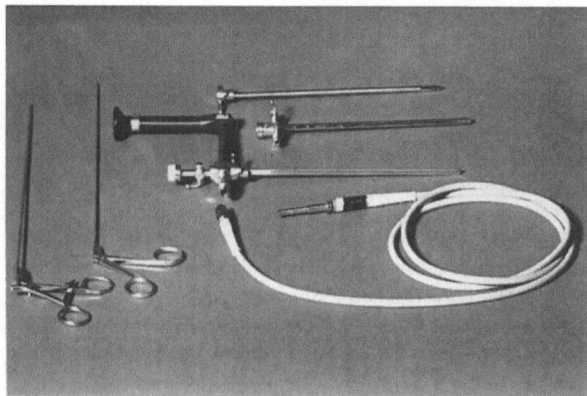

Fig. 6. Rigid endoscope and rigid instrumentation

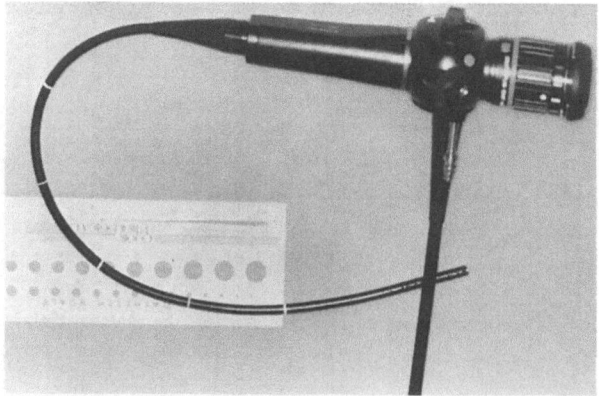

Fig. 7. Flexible endoscope of 4.8 millimeter outer diameter and two millimeter working channel

Procedure

By using a Mayfield adapter the patient is positioned in any desired position (supine, lateral, prone, etc), rotation as well as flexion-deflexion. After the patient's head is shaved at the entry level, the head and ring are completely cover with steril drapes; at the level of the ring where the localizing unit will be mounted, it is advisable to use transparent sterile drape. From now on, the procedure will be performed under completely sterile conditions which is fundamental in the management of cystic intracranial or intraventricular lesions. The sterile localizing device is mounted.

Under local anaesthesia, a burrhole is performed on the image defined entry area (transposed by the two calculated angles of the device). The dura is incised and coagulated in a cruciate fashion. The 2 mm stereotactic probe is inserted up to the desired depth (usually the center of a cystic lesion) to obtain diagnostic tissue and fluid for cytologic or microbiologic studies. After removing the probe, the metalic cannula (6 mm) with an obturator is inserted following the same trajectory through the adapter connected to the stereotactic arc. The rigid endoscopic optics (Fig. 6) is then introduced, and continuous suction and irrigation are connected. Irrigation is performed with 37 centrigrade saline solution in 10–13 cm of H_2O. The fluid component of lesions such as hematomas, abscesses, cystic tumours, etc., are then easily evacuated at the same time. Continuous irrigation keeps the optical system clean. Through the working channel, biopsy forceps can be used to obtain diagnostic tissue, scissors can be used to incise septations, and the Nd-YAG laser fiber can be used for coagulation, incision, or vaporization. In some cases of large cystic or intraventricular process, the use of the flexible endoscope provides further capabilities to inspect, resect, or vaporize tissue in otherwise inaccessible lesions. This is accomplished by tip flexion-deflexion and rotation of the endoscope. With flexible instrumentation, complete control of tumour cavities can be achieved and the Nd-YAG laser fiber can be used for further hemostasis or resection (Fig. 7).

Nd-YAG Laser and Endoscopy

The Nd-YAG laser has been selected mainly because of its properties of transmission through fiberoptics and its poor absorption in water characteristics, thereby enabling its use through the 2 mm channel of either a rigid or flexible endoscope by means of a 1.8 mm diameter disposable fiber (SLT, Malvern, PA). Cutting, hemostatis or vaporization can be achieved even under continuous irrigation and suction or in intraventricular lesions. Usually between 5 to 50 watts are used to achieve the specifically desired tissue effect. The Nd-YAG laser can be used in the conventional non contact technique or with contact probes adapted to the fiber tip.

The Nd-YAG laser fiber makes it possible to operate on deep lesions without mechanical contact through the small "surgical corridor" provided by the endoscope, thus reducing the need for manipulating and retracting the surrounding healthy brain tissue. The primary advantage of the laser, however, is that very limited lesions, whose extension and depths are predictable with a fair degree of accuracy, can be produced. Experimental data have shown that lesion volume can be increased by thermal spreading, which is related to the amount of energy delivered and more precisely to the time of irradiation. Tissue damage can be markedly reduced by using a very short exposure time[1]. Therefore, high power density and long irradiation time can be used to produce larger lesions. This is suitable for tumour resection when operating a distance from the healthy tissue, i.e., within the tumour mass. High power output and short irradiation time, however, limit the damage to the surrounding structure and can be used to obtain a sharp cutting or a more selective dissection, i.e, tumour margins. Short intermittent pulses at low power, focused upon blood vessels, can be used to produce safe coagulation within the brain tissue. Different laser–tissue interactions can be achieved by changing power, power density, time of irradiation, beam defocusing, etc. Depending on the specific situation, different effects are desired.

Cutting: Because of their high scattering, Nd-YAG lasers, which produce a wide, faintly wedge-shaped incision, are not as suitable for cutting as CO_2 lasers. Nevertheless, by adding contact tips, the YAG laser is turned into a cutting tool that produces a sharp incision (similar to that produced by the CO_2) and a more complete haemostasis. In this case, power ranges from 5–10 Watts and radiation are delivered in a continuous wave.

Vaporization and tissue removal: Most energy of the Nd-YAG laser is absorbed directly by nonwater tissue constituents to produce complete coagulative necrosis and haemostatis. The short pulse of 2–3 seconds at 40–50 Watts achieves a necrosis depth of 3–4 mm. By increasing the time of irradiation or the power density, temperature increases up to 100 centigrades and carbonization occurs with consequent enhanced surface absorption, which prevents further lesion depths. Because of the high thermal spreading, however, lateral extension of the damage increases. Greater lesion depths are therefore achieved by using separate pulses of 2–3 seconds with cooling periods of 20–20 seconds.

Haemostasis: The Nd-YAG laser can produce instantaneous occlusion of arteries up to 1–1.5 mm in diameter and veins up to 5 mm because of the thin wall and the reduced flow volume. Occlusion of larger arteries with thicker walls and high flow volume can be achieved with a defocused beam until spot size is at least three times larger than the vessel to be irradiated. Short intermittent

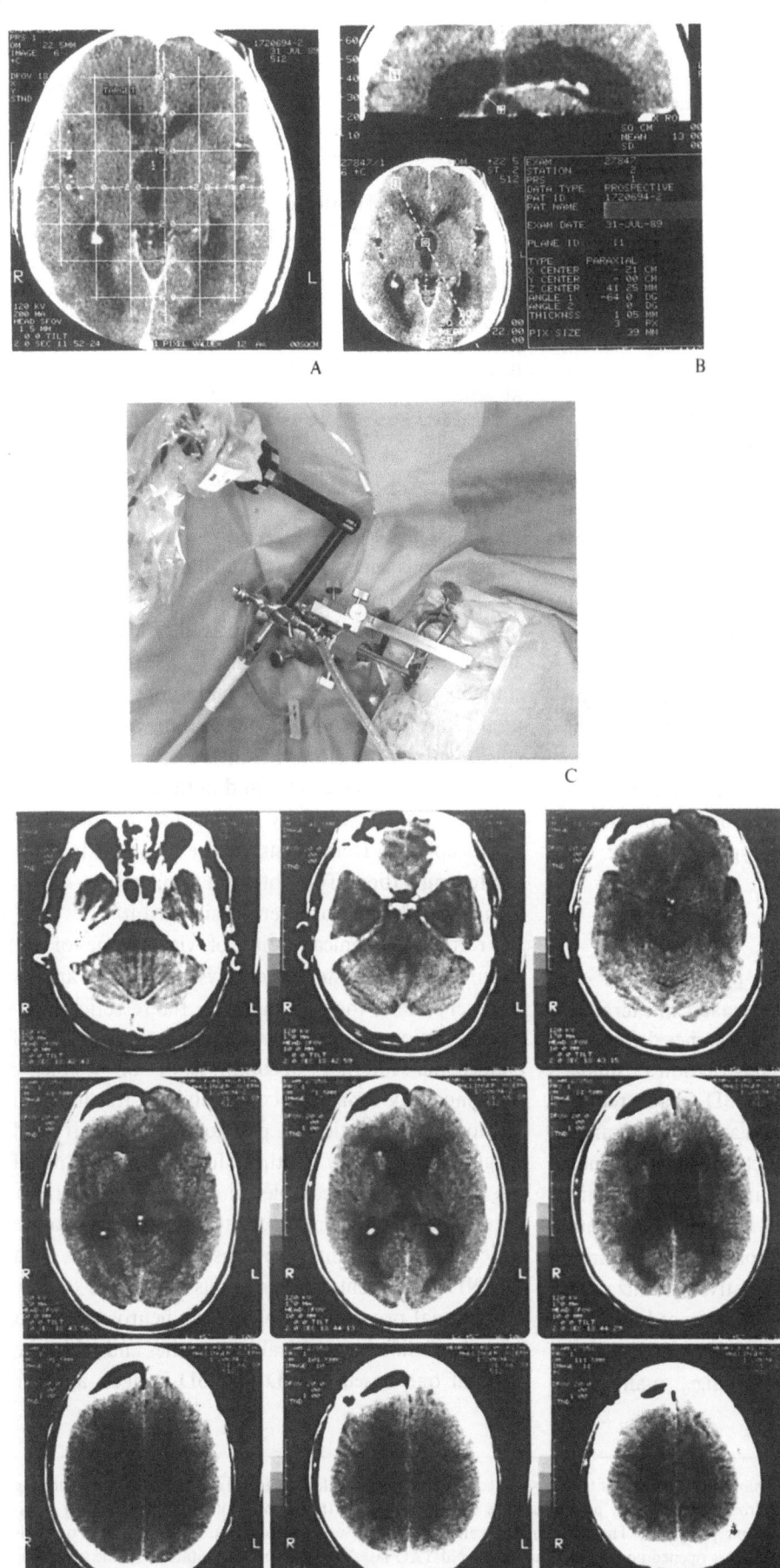

Fig. 8 (legend see page 74)

pulses of less than 3 seconds (to limit thermal spreading to the brain) at incident power of 40–80 Watts can be used. Another possibility, especially if working on a safe area of the brain, is to use a defocused laser beam delivered in a continuous mode for 10–15 seconds at 2–5 Watts.

Laser-tumour interaction: The thermal spreading ensuing from resection may produce delayed biological effects. Temperatures of 40–43 centigrades are present around the laser site. Because of the greater sensitivity to heat of malignant cells, this moderate hyperthermia may cause cellular damage and a local inflammatory process[2].

Clinical Experience

Our initial clinical experience with ELS on 76 patients has been encouraging. The technique has proved to be accurate, safe, and with low morbidity. Examples are shown in Figs. 8 and 9. Indications have been mainly cystic intraaxial (tumours, hematomas, benign cysts, abscesses, etc.) and some intraventricular lesions. Histological diagnosis has been obtained in all cases as well as in mediate internal decompression (Table 2). Operative mortality was seen in one patient (1.3%). This patient had an uneventful decompression of a large cystic metastatic adenocarcinoma of the lung, but respiratory arrest occurred secondary to probable intraoperative seizure of hypoxia. He was intubated and ventilated but died 3 days postoperatively due to irreversible brain hypoxic damage.

Discussion

Image-guided stereotaxis has proved to be an accurate and safe method to approach intracranial lesions, directing therapy to target volumes defined in 2D and 3D using computer reconstruction of image data[3,12]. With 2D-multiplanar or 3D image reconstructions, size, shape, and the main axis of intracranial lesions can be determined to select the best approach for diagnostic and therapeutic purposes[5,7,11,12]. This volumetric information can be transposed into stereotactic space. Integration of endoscopy into stereotaxis allows direct intraoperative visualization and monitoring of the stereotactic procedures[8,12]. Main indications of ELS technique include internal decompression of cystic tumours (i.e., malignant

Table 2. *Endoscopic ND-YAG Laser Stereotaxis (ELS): Indications and Diagnosis on 76 Patients*

Tumours	53
Glioblastoma multiforme	19
Anaplastic astrocytoma	8
Astrocytoma	4
Mixed oligo-astrocytoma	2
Craniopharyngioma	5
Metastasis	13
Plexus papilloma	1
Pinealoma	1
Cryptic AVM	1
Radionecrosis	4
Hematoma	8
Abscesses	5
Arachnoidal cysts	4
AIR	1

gliomas, metastases, craniopharyngiomas), benign cystic lesions such as arachnoidal cysts, colloid cysts, abscesses, and intracerebral hemotomas; other important indications for ELS are intraventricular processes, such as intraventricular haemorrhage, tumours, and some hydrocephalus. Flexible endoscopes can also be used as an isolated instrument in the management of epi- or subdural collections, such as chronic subdural hematomas, abscesses, etc., and in the management of intraventricular pathology such as haemorrhage and hydrocephalus. Hemostasis can be achieved by laser photocoagulation. The potential applications of this technology for neurosurgery is significant.

From the technical point of view, some of the features of the ZD stereotactic system make it appear as a very useful tool especially on the resection (centered craniotomy or microsurgical approach) and endoscopic internal decompression of lesions[12]: multimodality imaging compatibility, freedom on intraoperative patient's positioning, unobstructed surgical approach by multiple alternative mounting of the aiming device, complete intraoperative sterility, accurate target and trajectory transposition. Also, we believe the use of image processing capabilities, such as SNPS, represents an enormous advantage on any intracranial procedure allowing the neurosurgeon to "simulate" different surgical approaches and select the optimal one based on 2D and 3D images; another

Fig. 8. A) Axial view showing a cystic lesion on the third ventricule with secondary hydrocephalus. B) Paraaxial reconstruction showing the cystic lesion and the secondary hydrocephalus. The 2 points indicate the trajectory for the cannula of the endoscope. C) Intraoperative view of the stereotactic endoscopic procedure. The cannula is advanced up to the wall of the cystic lesion, then the optics is placed and under direct visualization the wall is opened and removed using the Nd-YAG laser and rigid instrumentation. At the same time the cavity is entered and the fluid content of a craniopharyngioma was removed. D) Immediate postoperative CT scan showing the decompression of the fluid content as well as partial removal of the cystic wall. The reduction of the secondary hydrocephalus is noticeable

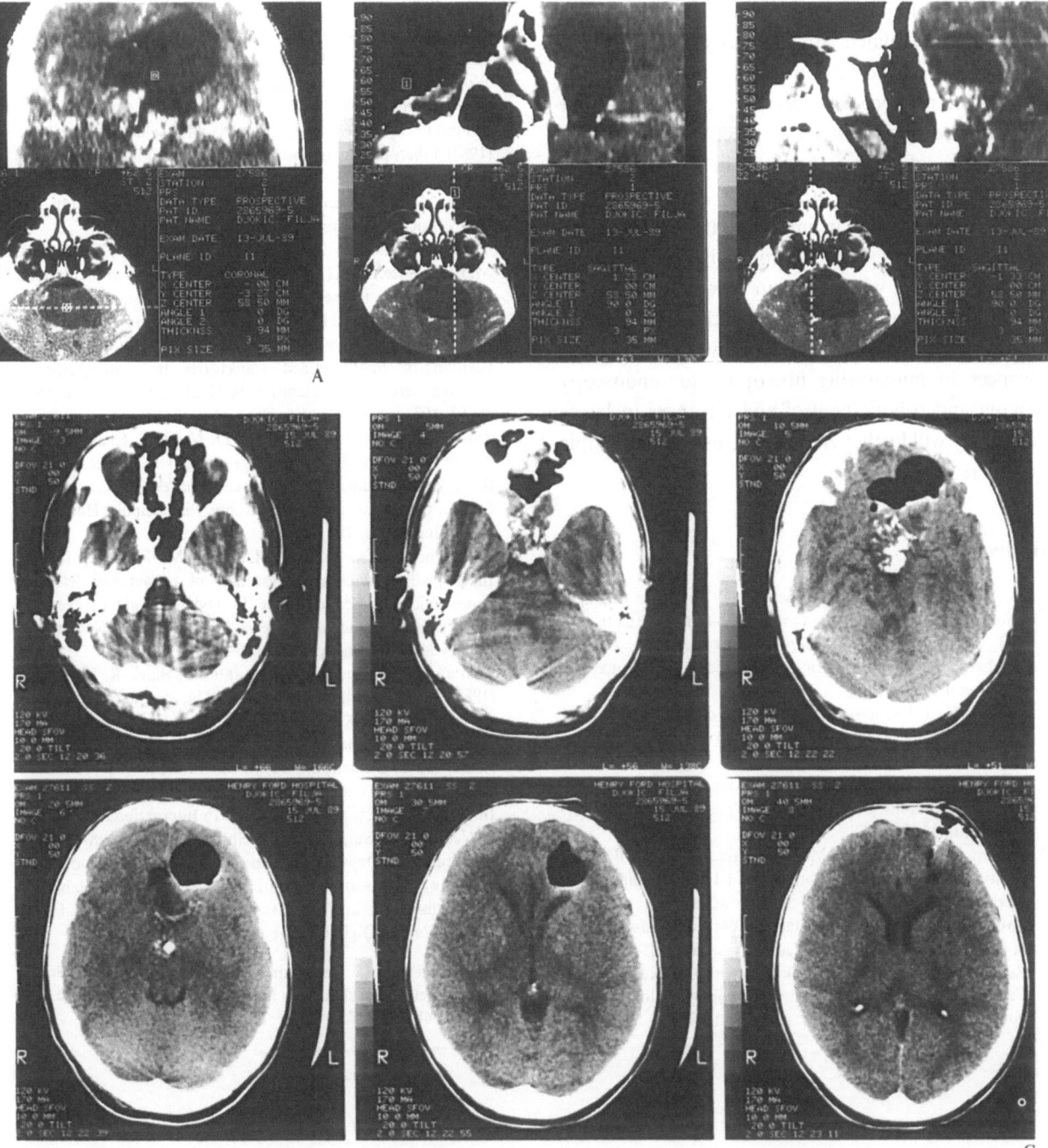

Fig. 9. A) Axial view and coronal reconstruction of cystic lesion on a patient that presented with acute blindness. B) Sagittal reconstruction of two different levels of the cystic lesions. C) Immediate postoperative CT scan following stereotactic endoscopic approach to the lesion. The fluid contents was completely removed and we can see rests of the solid component of this craniopharyngioma. The patient improved her visual status to a bitemporal hemianopsia

advantage is that once a plan is selected this can be accurately transposed into the operating room and there is a "preknown" anatomy at any specific depth of the selected surgical trajectory. Ultimately, all this represent an initial step on the implementation of automation in neurosurgical procedures.

From a general point of view, ELS appears to be a promising technique in the resection (cytoreduction) of some image-defined intracranial tumours and may be valuable in the management of malignant gliomas[5,9,11,12]. The impact of this therapeutic model as adjuvant to other volume-depending treatment modalities, such as brachytherapy, hyperthermia, phototherapy, etc., needs to be assesses[2,4]. Integration of image-guided lasers to control intracranial tumors allows precise removal and vaporization of these lesions[1]. The impact of this technology in survival needs to be evaluated. However, beyond the precise "cutting" or "vaporization" effect on tissues, and beyond the photocoagulator effect, both responsible for lasers being considered as precise surgical tools, applications of laser technology in stereotaxis will probably be derived from new diagnostic and therapeutic advances in image-guidance. The combination of laserbeam–tumour interaction and image guidance will result in new modalities of treatment for tumours, such as photodynamic therapy, and photothermally induced hyperthermia[2,4]. Future advances in miniaturing fiberoptics for endoscopy imaging and robotic control and advances in lasers technology will further enhance the use of endoscopic laser stereotaxis.

References

1. Beck O, Wilske J, Schonberger G (1979) Tissue changes following application of lasers to the rabbit brain. Results with CO_2 and Nd YAG lasers. Neurosurg Rev 1: 31–26

2. Bleehan N (1982) Hyperthermia in the treatment of cancer. Br J Cancer 45: 96–100

3. Kelly P, Kall B, Goerss S (1986) Results of computer-assisted stereotactic laser resection of deep-seated intracranial lesions. Mayo Clin Proc 61: 20–27

4. Mang T, Dougherty T (1986) Use of the Nd-YAG laser for hyperthermia induction as an adjunct to photodynamic therapy. Laser Surg Med 6: 237

5. Zamorano L, Dujovny M, Malik G, Yakar D, Mehta B (1987) Multiplanar CT-guided stereotaxis and I-125 interstitial radiotherapy: image-guided tumor volume assessment, planning, dosimetric calculations, stereotactic biopsy and implantation of removable catheters. Appl Neurophysiol 50: 281–286

6. Zamorano L, Dujovny M, Malik G, Mehta B, Yakar D (1987) Factors affecting measurements in computed-tomography guided stereotactic procedures. Appl Neurophysiol 50: 53–56

7. Zamorano L, Dujovny M, Malik G, Mehta B (1988) MRI stereotaxis: an indirect method of transposition of MRI-generated data into CT multiplanar images. Proceedings of the World Congress on Medical Physics and Biomedical Engineering 33: 40

8. Zamorano L, Chavantes C, Dujovny M, Malik G, Ausman J (1988) Fiberoptic neuroendoscopy and Nd-YAG laser interfaced to imaging-guided stereotaxis: a technical note. Presented at the Annual Meeting of the American Association of Neurological Surgeons, April 24–28

9. Yakar D, Zamorano L, Dujovny M, Cooper M, Grandhe R, Martin F, Sheehan M (1989) Interstitial temporary implantation of high activity iodine-125 sources for malignant glioma and brain metastases. Int J Radiat Oncol Biol Phys [Suppl] 17: 228

10. Zamorano L, Martinez-Coll A, Dujovny M (1989) Transposition of image-defined trajectories into arc-quadrant centered stereotactic systems. Acta Neurochir (Wien) [Suppl] 46: 95–103

11. Zamorano L, Dujovny M, Yakar D, Malik G, Chavantes C, Mehta B (1989) Multiplanar image-guided stereotactic brachytherapy with Iodine 125. In: Dyke et al (eds) Neurosurgery: State of the Art Reviews, Vol 4, Suppl. Philadelphia, Hanley & Belfus, Inc, pp 95–103

12. Zamorano L, Chavantes C, Dujovny M, Malik G (1989) Image-guided stereotactic resection of intracranial lesions: endoscopic and laser technique. In: Dyke P et al (eds) Neurosurgery: State of the Art Reviews, Vol 4, Suppl. Philadelphia, Hanley & Belfus, Inc, pp 105–118

13. Zamorano L, Dujovny M, Chavantes C, Block R, Flynn M (1989) Z-D multipurpose neurosurgical image-guided localizing unit. Presented at X Meeting of the World Society for Stereotactic and Functional Neurosurgery, October 2–5

14. Zamorano L, Dujovny M, Flynn M, Ausman J (1989) 3D-2D multiplanar workstation for neurosurgical stereotactic planning. In: Bhatia R et al (eds) Presented at 9th International Congress of Neurological Surgery, New Delhi, India, October 8–13

15. Zamorano L, Dujovny M, Ausman J (1989) Three-dimensional/two-dimensional multiplanar stereotactic planning system: hardware and software configuration. SPIE Vol 1153, Applications of Digital Image Processing XII: 552–567

Correspondence: L. Zamorano, M.D., Ph.D., Wayne State University, Neurological Surgery Department, 4201 St. Antoine 6E, Detroit, MI 48201, U.S.A.

Acta Neurochirurgica, Suppl. 54, 77–82 (1992)

Tumour Resection by Stereotactic Laser Endoscopy

F. Hor[1], M. Desgeorges[1], and **G. L. Rosseau[2]**

[1]Department of Neurosurgery, Val de Grâce Hospital, Paris, France, [2]Department of Neurological Surgery, George Washington University, Washington, D.C., U.S.A.

Summary

Recent advances in neuro-imaging have led to the early diagnosis of increasingly smaller and more deeply-seated tumours. Conventional neurosurgical techniques are often not satisfactory to deal with these lesions. The authors describe their preliminary experience with a prototype neuro-endoscope. The technical characterisstics of the instrument and description of its use in performing stereotactic laser tumour resection are provided.

Keywords: Stereotactic neuro-endoscopy; laser; MRI-guided stereotaxis; brain tumour.

Introduction

Recently, neuroendoscopes have appeared which permit therapeutic procedures[1,10] under direct vision: needle aspiration of cysts, lysis of membranes using microscissors and laser-guided tumour coagulation and removal. Thus appeared a new neurosurgical technique: neuro-endoscopy.

Our objective is to treat deep intra-ventricular or intra-parenchymal lesions using a neuro-endoscope under stereotactic guidance. We believe that the treatment of primitive, infiltrating cerebral tumours will develop toward a compromise between partial surgical removal and adjunctive therapy, such as photochemotherapy or thermotherapy.

Clinical Material and Methods

Endoscopic Material

Our endoscope has the following characteristic (prototype neuroscope: FORT Medical, 91410 Dourdan, France):

- 8 mm external diameter,
- 110° direct vision
- 2.2 mm working canal, permitting the passage of micro-instruments,
- Fiberoptic lighting system,
- Suction-irrigation system,
- Passage for 200, 400 and 600 µm laser fibers,
- 190 mm working length,
- Monocular direct vision and video control by camera CCD (KC 23.4213; FORT Medical) and video monitor (TM 90 PSN; JVC; Yokohama, Japan).

The use of this technique is justified, in our opinion, if the neuroscope is used under stereotactic guidance. We believe, in the near future, the development of medical imaging will be such that all cerebral surgical approaches will use stereotactic control. This may be laser-guided craniotomy and conventional microsurgery of superficial lesions or a neuroendoscopic removal of deeper lesions. The neuroendoscope is thus used like a biopsy needle which combines diagnostic efficiency with the therapeutic advantages provided by microinstruments and lasers.

The originality of the neuroendoscopes lies in its two systems (360° rotation on its axis (Fig. 1) and, distally, 90° angulation of the laser fiber (Fig. 2). In fact, we consider it as important for the neuroendoscope to be locked to the frame and immobilized; two graduated cylinders direct the rotation of the assembled apparatus. Only the working canal is fixed, since it is in the center of the neuroendoscope. One can imagine the combination of two movements (rotation and distal angulation) which permit the laser-beam to describe a hemisphere in which the tumour is situated. Thus, it is important to precisely calculate the position of the neuroendoscope in relation to the tumour mass.

Results

Determination of Neuroscopic Trajectories

The size of the instrument requires that one determines precisely the parenchymal structures along the trajectory.

The acquisition of the primary stereotactic images is performed under local anaesthesia just before the neuroscopic operation (Fig. 3). MRI is routinely used (Signa G. E, 1.5T). After initial images are obtained in 3-D, a paramagnetic contrast agent (Gd-DTPA) is injected to allow visualization of the target and calculation of co-ordinates (Fig. 4).

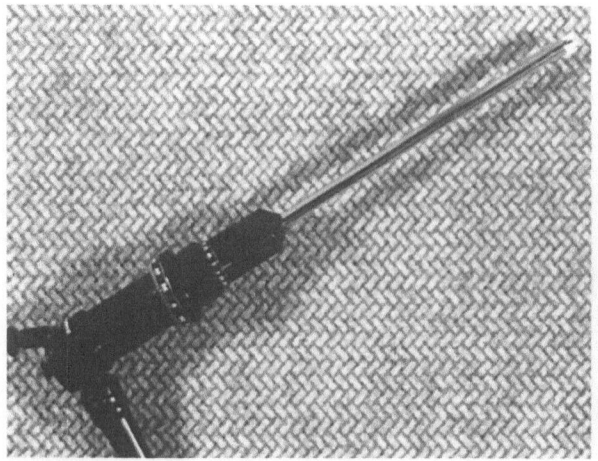

Fig. 1. Neuroendoscope (FORT Medical prototype)

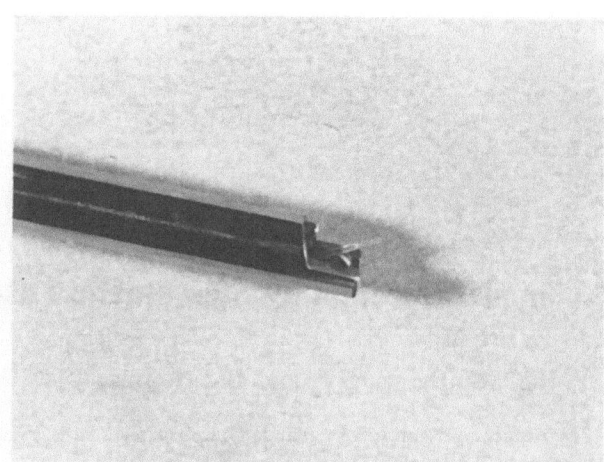

Fig. 2. Neuroendoscope, distal extremity showing the 40° angle of the YAG laser fiber

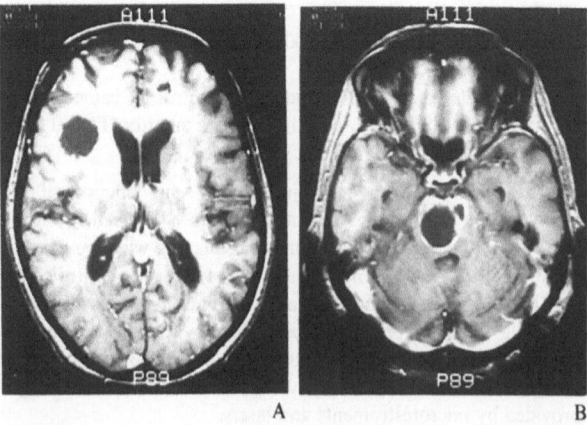

Fig. 3 A, B. MRI performed in the stereotactic frame showing evidence of two voluminous cystic lesions, in the right frontal (A) and bulbo-peduncular regions (B)

Fig. 3 C, D. Same patient, contrasted CT scan at the bulbo-peduncular level: (C) before and (D) after the stereotactic procedure

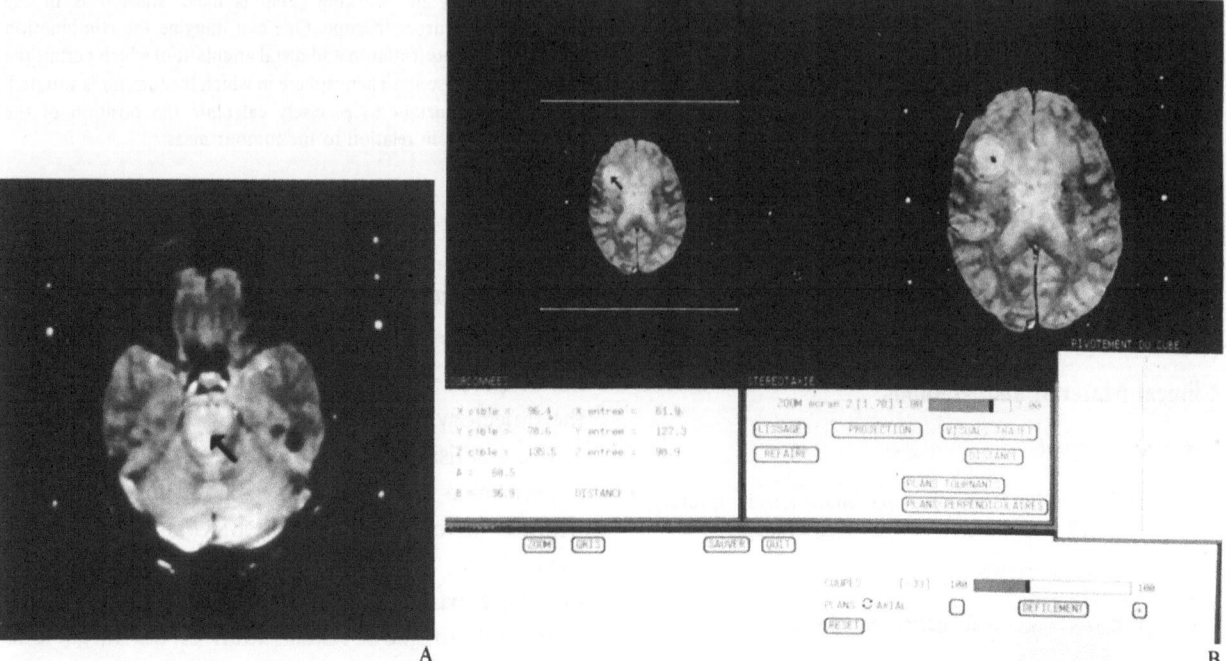

Fig. 4. A) Determination of the distal bulbo-peduncular target (arrow). B) Determination of the proximal right frontal target (arrow). The X, Y and Z coordinates of the 2 targets as well as the angles A and B are automatically displayed at the stereotactic work station

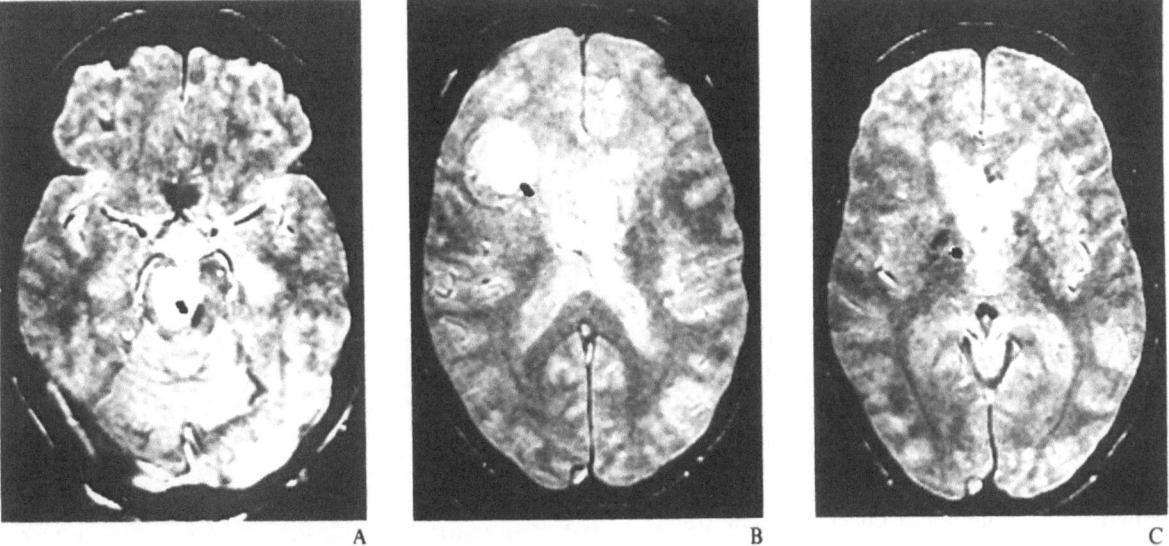

Fig. 5. Axial 3D-MRI depiction of the trajectory and thus of the position of the biopsy needle or neuroendoscope

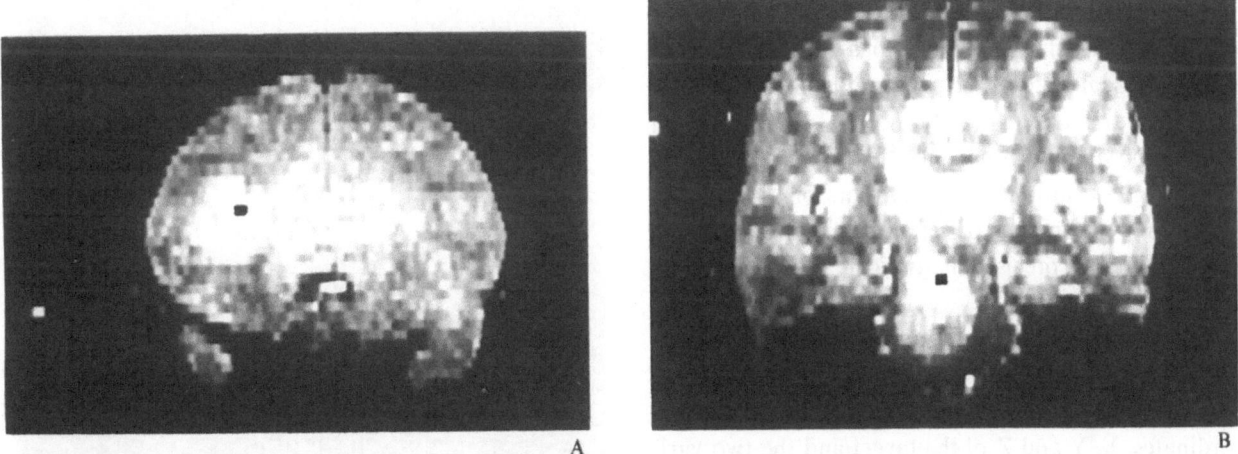

Fig. 6. Trajectory control scan in 3D-MRI reconstruction, coronal cuts. A) At the level of the right frontal cyst. B) At the level of the bulbo-peduncular cyst

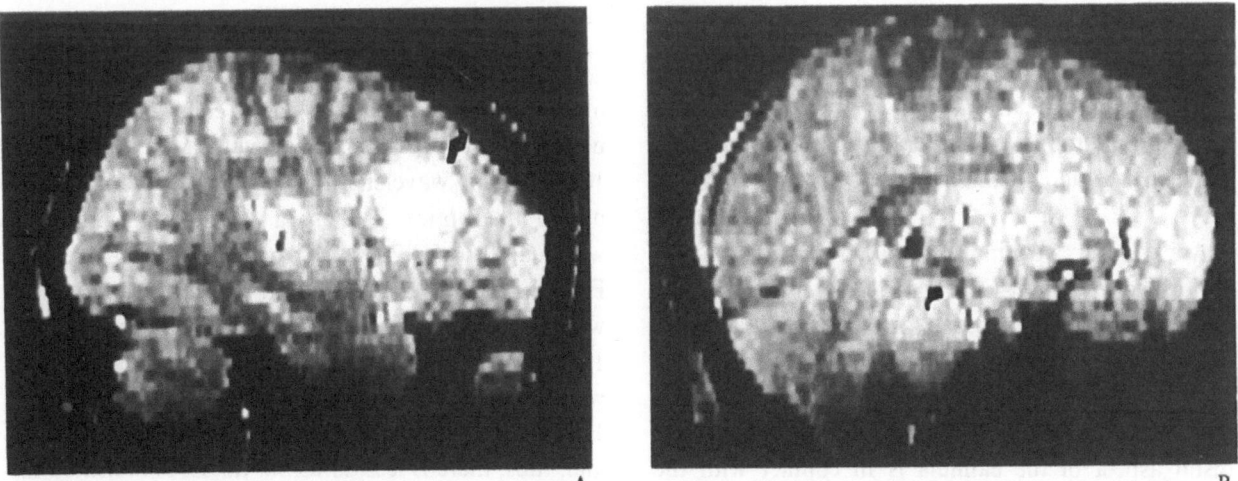

Fig. 7. Trajectory control scan, 3D-MRI reconstruction, sagittal cuts. A) At the pre-tumoural level, deep to the cerebral cortex. B) At the bulbo-peduncular junction

The imaging data are assembled at the SUN 3.2 stereotactic work station. The program allows visualization along the axial, coronal and sagittal planes under stereotactic conditions. Two methods for trajectory calculation are possible: one may either choose two targets, or one target and the angle of the trajectory. In either case, the results are the determination of a unique trajectory.

This trajectory is then visualized in 3 planes in the form of a point which exactly represents the position of the neuroscope. (Figs. 5, 6, 7). This also permits a check against passing too close to any vascular structure. If it is determined that a modification of the trajectory is necessary, one need only change the one or two target points and check the new trajectory in the planes. Finally, the program gives the target-cortex distance, the target defined as the first encounter of tumour by the endoscope once it is put in place.

It is also possible, using the stereotactic work station, to visualize the planes passed by the axis of the neuroscope (rotating planes, Fig. 8) and planes perpendicular to this axis. Finally, 3-D exploration combining simultaneous axial and sagittal cuts (Fig. 9) gives a good notion of the contours of the tumour. These new techniques of cerebral imaging allow the recognition of precise anatomical structures along planes.

Once the trajectory is approved, the intended position of the neuroscope is determined in the isocentric stereotactic system by 5 values: the 3 co-ordinates, X, Y and Z of the target and the two variable angles in the antero-posterior and transverse plane.

Placement of the Neuroscope

The stereotactic frame remains in place. Under general anaesthesia, a 2 cm craniotomy is performed which is centered on the pre-determined trajectory.

After dural opening, a limited corticotomy is performed. A blunt-tipped stylet is introduced to correspond the target-cortex distance. A rigid plastic cannula adapted to the diameter of the endoscope is slid over the stylet; the length of this cannula is also determined by the target-cortex distance. Thus, the distal aspect of the cannula is in contact with the tumour. The stylet is removed and the neuroscope is inserted via the cannula.

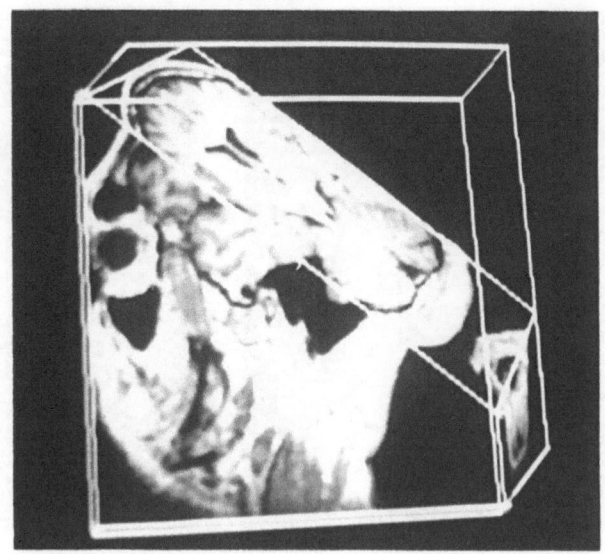

Fig. 8. "Rotating" planes provided by the stereotactic work station

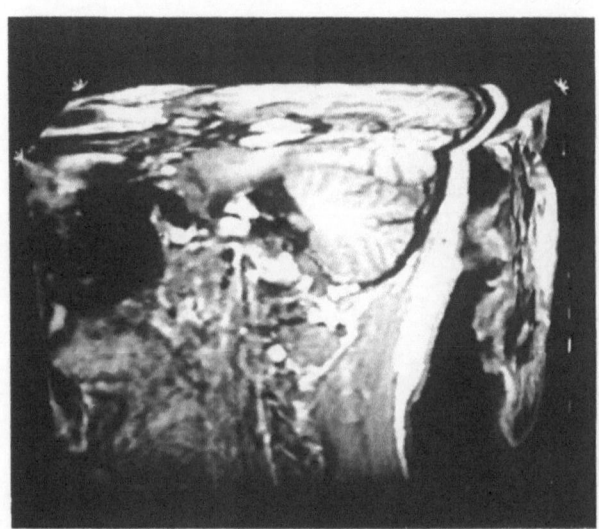

Fig. 9. Simultaneous "axial-sagittal" exploration

Laser Tumour Resection

Any laser coupled with a neuroscope must have a wavelength compatible with the employed use of fiberoptic cables. The CO_2 laser, commonly used in neurosurgery, uses nanawaves with a wavelength of $10.6 \mu m$. This wavelength is not compatible with the use of fiberoptics. We use a YAG laser with two wavelengths (MC 2100, Microcontrole; 91005 Cedex; Evry, France) 1.06 and $1.32 \mu m$ which are very useful with the neurosurgical endoscope. In fact, the laser of $1.06 \mu m$ wavelength diffuses into the tissue and is well-suited for the coagulation of small vessels. The 1, $1.32 \mu m$ wavelength is absorbed by the tissue fluid and, thus, there is less diffusion. Its use is indicated in the treatment of tumour which is close to normal cerebral parenchyma.

Clinical Results: (Preliminary Experience)

Two patients have been operated on under neuro-endoscopic conditions.

One presented with colloid cyst of third ventricle resected under endoscopic control; the other affected with hydrocephalus has been treated by opening of lamina terminalis and laser plexectomy.

Discussion

Tumour vaporization is carried out under direct vision, using a 400 μm fiber, 1.06 μm wavelength and 40–50 watt superpulsed mode. An intratumoural cavity is thus created, allowing the use of a 200 mm laser fiber for vaporization of the lateral tumour mass. During the procedure, smoke and debris are evacuated via the suction or working canals.

Regular irrigation with isotonic saline is performed, which allows cleaning of the optics and removal of debris from the tumour cavity.

The laser is programmed differently when using a wavelength of 1.32 μm. With the power at 25 W, a "superpulsed" mode is preferred. The continuous mode appears to provide more efficient tumour removal, but the temperature elevation in the irradiated tissue precludes its use in certain cases. Thus, in liquid environments, so that the laser beam need not pass through structures filled with fluid, such as serum or CSF, or functional parenchyma. The absorption of thermal energy by such fluids causes them to rapidly come to the boil. It is always possible to surmount this difficulty by only using the laser fiber when it is in contact with the tumour and never from a distance. But there remains a risk of burning the distal aspect of the fiber, rendering it useless.

Haemostasis

If the tumour resection results in a small amount of bleeding, irrigation with cold saline is necessary. This cleanses the objectives and improves vision. For vessel coagulation, the laser settings are different: 1.06 μm wavelength, low power and continuous mode. If haemostatic control of a vessel is not possible, one must use bipolar coagulation via the operator canal (Ultrafine bipolar forceps, F. L. Fischer).

Temperature Control

A thermosound is used to control the temperature in the tumour cavity and the neighbouring normal parenchyma. During tumour vaporization, the temperature 2 mm from the point of laser impact can reach 120° C (wavelength: 1.06 μm, power: 50 W, continuous mode) and only 60° C (wavelength: 1.32 μm; power: 35 W; super-pulsed mode).

Photochemotherapy

Photochemotherapy is a recently developed technique which is used infrequently in neurosurgery. It is based on the principle of specific photosensitivity of tumour cells. Under the influence of a specific type of radiation, certain molecules in the tumour cell become cytotoxic. This attractive technique approximates surgery at the "cellular" level. Most authors[7,9] have published work on haemotoporphyrin derivates (HPD). We are performing in vito studies on two photosensitive products: a porphyrin (photofrin II, Lederlee, France) and a phthalocyanin (TINOLUX BBS derivative, CIBA-Geigy, Switzerland).

We use a dye-laser (MC 600; MICROCONTROLE, Evry, France) which permits the regulation of radiation between 620 and 675 nm. Neuroendoscopy permits photodynamic therapy (PDT) treatment of the tumour margins after gross total resection.

This work will be published in the near future.

We are currently in the process of refining certain technical aspects of the neuroendoscope.

Conclusion

Improvements in neuro-imaging, and the resultant discovery of smaller lesions, require a response in the form of new neurosurgical instrumentation. It is useless to diagnose a deep 1 cm diameter lesion if we lack the means to remove this lesion. The challenge for our speciality in the near future will be to improve our therapeutic interventions while always preserving the functional capacity of the patient.

References

1. Appuzo M, Chandrasoma P, Breeze R, Cohen D, Luxton G, Mazumder A (1968) Applications of image-directed stereotactic surgery in the management of intracranial neoplasms. In: Heilbrun P, et al (eds) Concepts in neurosurgery, Vol. 2. Williams and Wilkins, pp 73–132
2. Auer LM, Holzer P, Ascher PW, Heppner F (1988) Endoscopic neurosurgery. Acta Neurochir (Wien) 90: 1–14
3. Bullard DE (1989) Intra-operative impedance monitoring during CT guided stereotactic biopsies. Stereotact Funct Neurosurg 52: 1–17

4. Fukushima T (1978) Endoscopy of Meckel's cave, cisterna magna and cerebellopontine angle. J Neurosurg 48: 302–306

5. Kelly PJ, Goerss SJ, Kall BA (1988) The stereotaxic retractor in computer-assisted stereotactic microsurgery. Technical note. J Neurosurg 60: 301–306

6. Powell MP, Torrens MJ, Thomas JLG, Horgan JG (1983) Isodense colloid cysts of the third ventricle. A diagnostic and therapeutic problem resolved by ventriculoscopy. Neurosurgery 13: 234–237

7. Powers SK (1988) Photochemotherapy. In: Cerullo LJ (eds) Application of lasers in neurosurgery. Year Book Medical Publishers, pp 137–155

8. Shelden CH, Mc Cann G, Jacques S, Lutes HR, Frazier RE, Katz R, Kuki R (1980) Development of a computerized microstereotaxic method for localisation and removal of minute CNS lesions under direct 3-D vision. J Neurosurg 52: 21–27

9 Wharen RE, Anderson RE, Laws ER (1988) Photoradiation therapy of malignant brain tumours. In: Cerullo LJ (eds) Application of lasers in neurosurgery. Year Book Medical Publishers, pp 156–171

10 Zamorano L, Chavantes C, Dujovny M, Malik G, Ausman JI (1989) Neuroendoscopic image-guided stereotactic. Acta Neurochir (Wien) 98 : 91

Correspondence: F. Hor, M. D., Department of Neurosurgery, Val de Grâce Hospital, 74 Bd de Port Royal, F-75230 Paris Cedex 05, France.

Acta Neurochirurgica, Suppl. 54, 83–88 (1992)

Volumetric Stereotactic Resection of Superficial and Deep Seated Intraaxial Brain Lesions

A. Camacho and **P.J. Kelly**

Department of Neurosurgery, Mayo Clinic, Rochester, Minnesota, U.S.A.

Summary

From 1984 to 1989 a total of 374 computer-assisted stereotactic resections based on computed tomography or magnetic resonance imaging were performed on 337 supratentorial and 37 infratentorial brain lesions. Computer-assisted stereotactic volumetric resection allows a more aggressive extirpation of tumours with less damaging of the adjacent brain tissue. This procedure is of most benefit in deep seated circumscribed lesions and of less benefit in infiltrating tumours such as high grade gliomas and in fibrillary astrocytomas located in essential brain areas.

Keywords: Intra-axial brain tumours; computer-assisted volumetric-stereotaxis; stereotactic laser resection; brain shift.

Introduction

The use of computer-assisted volumetric stereotaxis in neurosurgery has its major advantages in the resection of deep seated lesions. The reasons for this are threefold.

First, the surgeon's three dimensional orientation decreases below the cortical surface. In using standard craniotomy methods to resect tumours below the cortical surface there is a risk that the surgeon could get lost in an attempt to find the tumour.

Second, lesions are most often irregular in shape, and once found at conventional craniotomy, the surgeon can get lost in the various extensions of the tumour.

Third, the boundary between tumour and deep gray matter nuclei is frequently unclear, such that the surgeon may not be able to identify where tumour ends and normal brain begins. Thus, with computer providing a guide in three dimensional space, as our experience will show, deep seated lesions can be removed with low morbidity and mortality.

Methods

Computer reconstruction of stereotactic CT and MRI defined lesion contours allows the representation of a tumour volume in three dimensional or stereotactic space. This is retained within the computer matrix. The computer then shows the surgeon a series of slices which indicate the image defined boundaries of the tumour which are related to stereotactically placed surgical instruments which are in turn used to resect the tumour.

For deep seated lesions a carbon dioxid surgical laser is used. The procedure is useful for centrally located, superficial and deep-seated intracranial lesions.

The procedures are done in three essential steps:

1. The data base acquisition,
2. the surgical planning, and
3. the surgical procedure.

Stereotactic Headframe

The data base acquisition and the surgical procedure are performed in an imaging compatible stereotactic headframe which is placed under neuroleptic sedation and local anesthesia. The frame is fixed to the skull by means of fixed length flanged carbon fiber pins. The pins hold the frame rigidly to the skull, detachable micrometers are used to measure the extensions of the pin beyond the plane of vertical support. Osseous fixation holes and fixed length of the fiber pins along the micrometer measurements provide a mechanism by which the frame can be applied and reapplied at a later time in a precise manner.

1. Data Base Acquisition

CT-scanning: The first item on the data base is CT scanning. CT scanning is precise three dimensional data. If one gathers this data under stereotactic conditions, one can use this volumetric data in stereotactic space. Patients undergo CT scanning in the stereotactic headholder to which a localization system consisting on nine carbon fiber rods arranged in the shape of three "N" is applied. There is one on either side of the head and one anteriorly. The carbon fiber rods produce a series of nine reference marks around each CT slice, (Fig. 1). The vertical height of the slice is determined as follows. One can simply measure the distance between the reference mark made by the vertical from that made by the oblique.

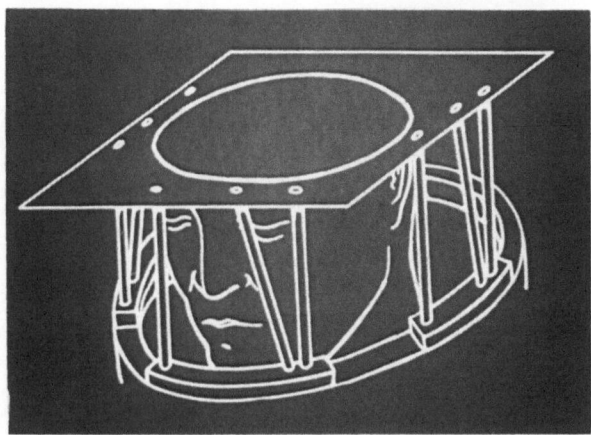

Fig. 1. CT Scan localization system. Cross-section of the "N's" by a plane in space creates a series of nine reference marks from which the computer derives the exact position of the plane above the stereotactic head frame basering

One has then one side of a right triangle for which the angle is known and thereby gets a vertical height in space above basering This results in the determination of three points is space which defines the orientation and location of a plane in space. These reference marks are used to derive stereotactic coordinates for any point on each CT slice.

MR-imaging: MR imaging is also precise three dimensional data. The procedure for stereotactic MRI is similar to that used for CT except that the "N" shaped localization systems consisting of capillary tubes filled with copper sulfate are used.

However, multiplanar imaging is possible with MRI. Therefore, the localization devices are on either side of the head, anteriorly, posteriorly and superiorly to allow imaging in the axial, coronal and sagittal plane.

Stereotactic angiography: In stereotactic intracranial procedures it is useful to know the presice three dimensional position of important cranial vessels. In adition, stereotactic angiography will also indicate the position of the major sulci and fissures. Stereotactic digital angiography is performed utilizing reference pellets embedded in a lucite plate located on either side of the head, anteriorly and posteriorly. This results in a series of 18 reference marks from which magnification and stereotactic coordinates can be determined.

Each angiogramm is performed in orthogonal and precise six degree stereoscopic pairs which, when viewed stereoscopically, allow one to reconstruct the vascular structure in three dimensional space. If one simply selects the deep segments of all the cortical arteries and veins one can with the help of a stereoscopic merge unit, determine the position of the deep sulci since the arteries and veins run on the surface of the gyri and deep into the sulci.

Data transfer: Following completion of the stereotactic CT, MRI, and digital angiography examinations, the data are transferred from imaging host computer to an operating room computer system. The computer consists of a Vicom image processor, a Sun 4 work station and a Data General 9800 backup computer. These are controlled locally and remotely by an image display panel in the induction room adjacent to the operating rooms. The computer generated image data are displayed on monitors which hang from ceiling in each of the two operating rooms equipped for stereotactic surgery. The rooms are also equiped with ap and lateral laser collimated fixed tube teleradiography.

2. Surgical Planning

In computer assisted stereotaxis we deal with a point in space for simple procedures such as stereotactic biopsy, third ventriculostomy, functional procedures and simple cyst aspiration.

However, if one is interested in therapeutic intervention such as stereotactic resection or planning interstitial irradiation, one must construct a volume in space. This is done as follows.

The computer displays sequential slices of the CT and MRI series on the consoles. It automatically recognizes the nine reference marks from the localization system and places these slices in a precise position within a three dimensional image matrix. The surgeon then traces around the CT defined limits of the tumour and around the MRI defined limits of the tumour, and the computer then stacks these slices in its three dimensional image matrix. Once these slices are stacked in space at precise intervals, the computer interpolates intermediate slices 1 mm intervals and fills each of these interpolated and digitized slices with 1 mm voxels which allows the establishment of a volume in space. The volume is established for CT scanning and MRI defined limits of the tumour. Finally, if the surgeon indicates to the computer the actual direction at which the tumour is going to be approached, i.e. gives the computer a viewline (angle from the horizontal plane and angle from the vertical plane) the computer will then slice and reformat that data, then slice the tumour volume perpendicular to the defined viewline (Fig. 2).

The viewline is very simply, the surgical approach expressed mathematical terms. Once the computer slices defined tumour along the viewline, it displays each of those sequential slices on a monitor in the operating room, showing the CT and MRI defined limits of the lesion portrays the stereotactic coordinates necessary to have this picture in the stereotactic field. These images are directly related to the stereotactic arc-quadrant frame in the operating room. These are displayed on a monitor in the operating room and in a "heads-up" visual display. The stereotactic frame is an arc-quadrant in which the patient's head is moved in X, Y and Z directions to place a target point into a focal point, or isocenter, of an arc-quadrant.

In the Compass Stereotactic System, each of the frame movements are executed by computer driven stepper motors in X,Y, and Z. Manual and electronic backups exist for each of these systems.

"Heads-up"-display: While performing microsurgery it is cumbersome to lift ones eyes from the operative field to reorient oneself

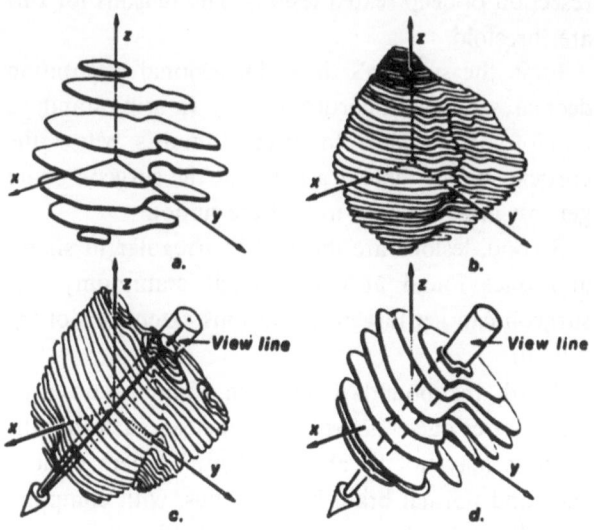

Fig. 2. Computer recreation of tumour volumes from CT or MRI stereotactic acquired data. The tumour is oriented along the view line

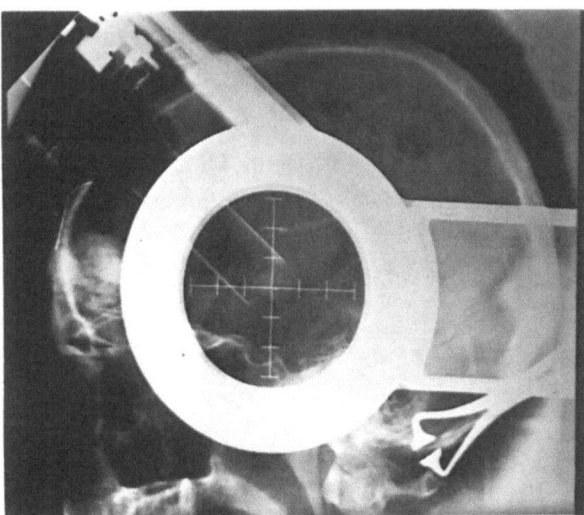

Fig. 3. Intraoperative radiograph showing the cylindrical stereotactic retractor advanced to the outer margin of the tumour

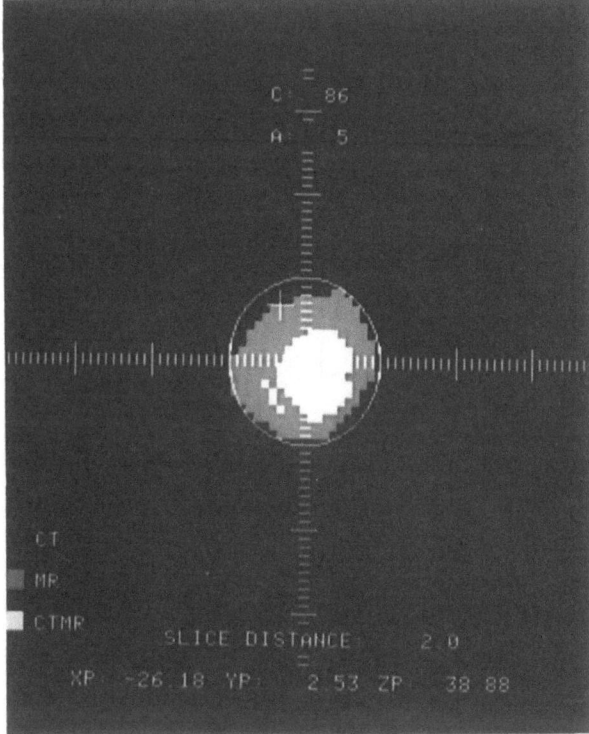

Fig. 4. Computer outline of the retractor with the MRI (outer) defined tumour limits, and the CT (inner) defined tumour limits, as viewed by the surgeon

by the computer generated image on the ceiling mounted monitors and then back to operative field. Therefore, based on military technology, we constructed a "head-up" display for an operating microscope similar to that seen on jet fighter aircraft in which the computer generated image is reflected into the surgeon's eye during the procedure and the actual computer generated image is the exact size of the surgical field and superimposed on it. The surgeon can continue to work without ever averting his eyes from the surgical field with the computer generated image proceeding a template around to facilitate identification of the plane between tumour and surrounding brain tissue.

3. Surgical Procedures

In order to remove an anterior thalamic tumour for example, the following would be the sequence of steps. One first makes a linear incision at the hairline and opens the skull with an $1-1\frac{1}{2}$ inch trephine. Then using the stereotactic frame and carbon dioxide laser, an incision is made through the cortex and subcortical white matter to the outer border of the tumour. At outer border of the tumour the incision is undercut somewhat. The computer by the way has calculated the range of the tumour along the viewline. A retractor is then advanced to the tumour. The retractor is cylindrical in shape and mounts on the stereotactic arc-quadrant. This creates a channel from the surface to the outer border of the tumour. The retractor is advanced by first making a 2 cm cortical incision and subcortical white matter using the laser (Fig. 3). At the outer border of the tumour the computer shows us the configuration of the retractor and the most superficial slices of the tumour. The computer shows us the outline of the retractor as viewed by the surgeon, shows the limit of the tumour defined by CT and by MRI at these given stereotactic coordinates at a given distance along the surgical viewline (Fig. 4). In practice, a surgeon works around the tumour to first separate it from the surrounding brain utilizing the laser and then vaporizes this slice by slice going from most superficial areas to the deepest.

This technology is also useful for removing superficial lesions. The lesion is built up as a volume and then a pilot hole is made directly over the lesion. Then utilizing only a linear incision, one can use this pilot hole to swing a trephine. The trephine itself has a known position in stereotactic space. The computer can show us an image of the trephine which is superimposed through the "head-up" display onto the actual surgical field. The computer

derived imaging based tumour slices are projected with respect to the location of the trephine and the surgeon can then use this outline a template to remove the tumour. One very simply resects the area that corresponds in appearence, location and shape to the actual computer generated image.

The procedure can also be used for posterior fossa resections, with a slight modification. First the headframe is applied in an inverted position for data acquisition and for the surgery. The patient is then operated prone in the stereotactic frame (Fig. 5).

Results

In a five year period between August 1984 and August 1989 a total of 374 various intracranial lesions were operated on with the described procedure. Of these, 337 lesions were supratentorial and 37 in the posterior fossa. For histology see Tables 1 and 2.

Metastatic Tumours

Perhaps the most straight forward use of the described technology is in the stereotactic resection of metastatic tumours. Metastatic tumours are well circumscribed lesions which can be totally resected. Figures 6 and 7 demonstrate an example which can certainly benefit by stereotactic removal: A renal cell carcinoma located just posterior to the central sulcus which was removed as a single specimen. Postoperatively the patient was neurologically intact.

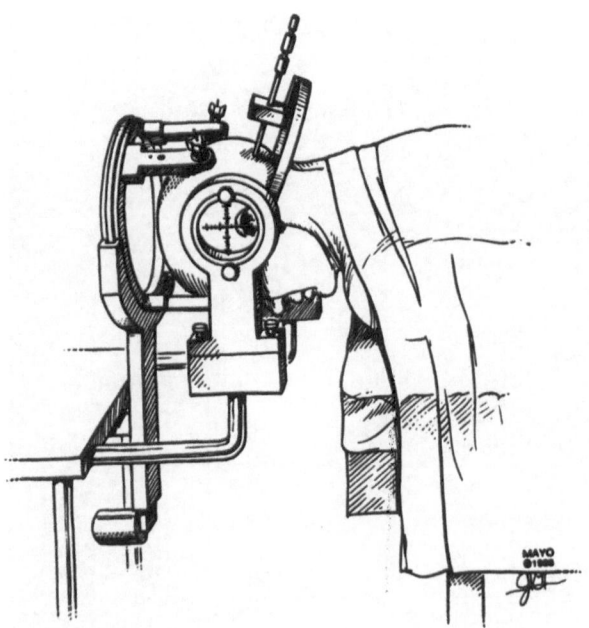

Fig. 5. Surgical configuration for a posterior fossa case. The head frame is applied in an inverted position and the patient is prone in the stereotactic headframe

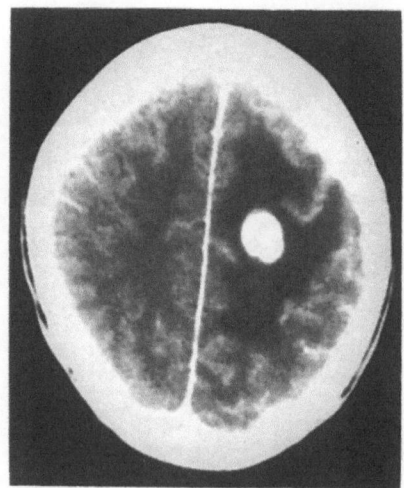

Fig. 6. CT scan (*left*) showing a metastatic renal cell carcinoma lesion just posterior to the central sulcus

Table 1. *Histology Neoplasms*

Astrocytomas	
Grade IV	71
Grade III	15
Grade II	15
Grade I	5
Pilocytic	45
Oligodendrogliomas	32
Oligo-astrocytomas	13
Ependymomas	4
Subependymomas	2
Neurocytoma	2
Medulloblastomas	1

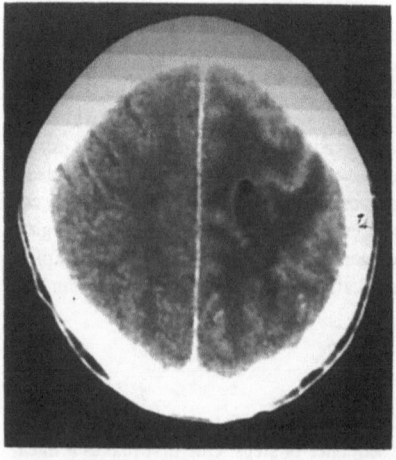

Fig. 7. Postoperative CT scan of the same patient as in Fig. 6, showing complete removal of the metastatic lesion

Table 2. *Histology Non-neoplastic*

Radiation necrosis	5
Gliosis (seizures)	5
Abscess	3
Tuberous sclerosis	4
Hematomas	2
Vascular lesions	36
Cystocercosis	1

Glioma

The tumours which have been treated most commonly were glial neoplasms. Using volumetric stereotactic methods we remove any part of the tumour, which is proposed to be removed. In high grade gliomas this is limited to the volume of the tumour defined by CT or MRI contrast enhancement. Figures 8 and 9 show a preoperative and postoperative scan of a grade 4 astrocytoma located in the left posterior superior thalamus. This was removed through the superior parietal lobule. Preoperatively the patient has a very mild right hemiparesis, which postoperatively was the only neurological deficit. Nevertheless, these patients with high grade gliomas are not cured, but their survival curve slightly expand.

Low-grade gliomas are basically of two types. There are fibrillary astrocytomas seen in adults which are characterized by an area of low density on CT, and areas of increased T1 and T2 on MRI. These are not lesions that can be resected in essential brain areas because they are comprised of isolated tumour cells within infiltrated parenchyma. Resecting these tumours involves resecting functioning brain paren-

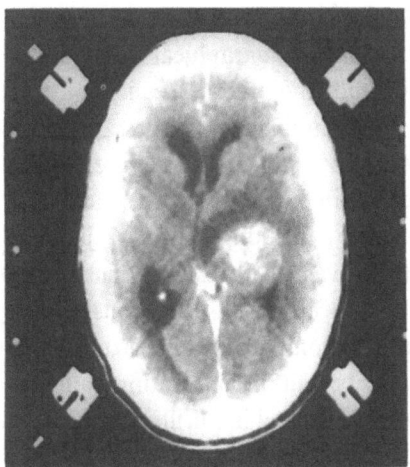

Fig. 8. CT scan showing grade 4 astrocytoma in the left posterior thalamus

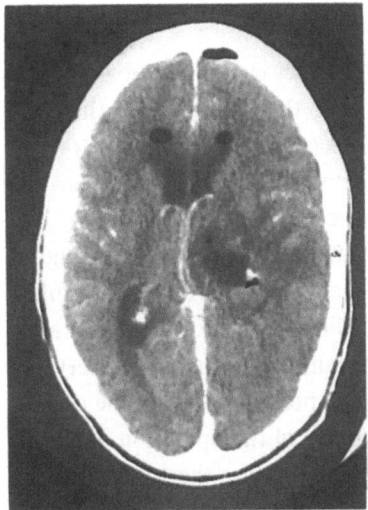

Fig. 9. Postoperative scan, same patient as in Fig. 8, showing complete removal of the contrast enhancing portion of the tumour

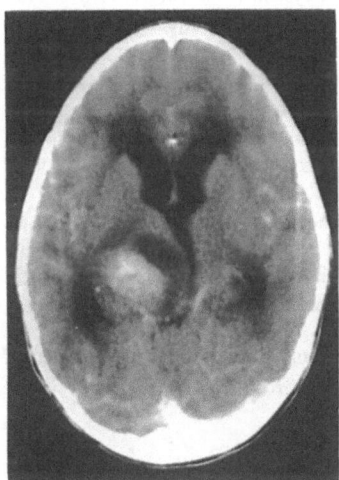

Fig. 10. CT scan of right posterior ventral thalamus pilocytic astrocytoma

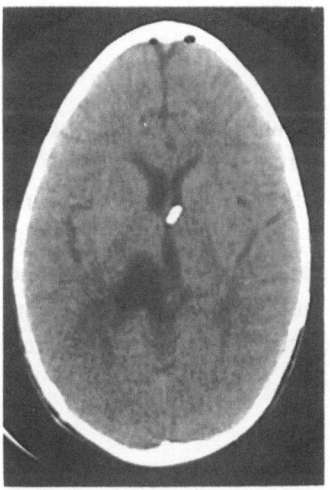

Fig. 11. One year postoperative CT scan showing complete removal and no evidence of recurrence

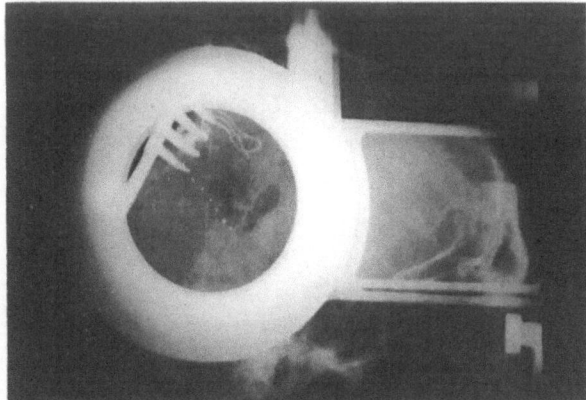

Fig. 12. Intraoperative radiograph showing a series of stainless steel balls deposited along the view line. The stainless steel balls serve as a reference for any shifts that might occur during the operative procedure (see text)

chyma. However, occasionally these tumours are located in nonessential brain areas in which case they can be resected. Often these lesions will present seizures.

Pilocytic astrocytomas are another type of low grade astrocytomas which can be benefitted by stereotactic resection. Figures 10 and 11 show a preoperative and postoperative scan of a 5-year-old-girl with a pilocytic astrocytoma in the posterior thalamus. The postoperative scan was done one year following surgery with no evidence of recurrence. Resection of pilocytic astrocytomas may well be curative. Our longest follow-up is a 15-year-old boy who at age 7 had a resection of a ventral thalamic pilocytic astrocytoma. Eight years after surgery CT scans show no evidence of recurrence.

Brain Shifting

A question that often arises is, how to know that the brain does not shift once the cranium and the dura have been opened? To detect any such shifts, especially in cases of cystic tumours or tumours near the ventricular system, what is done first is to tap the tumour with a stereotactic probe directed through a 1/8 inch twist drill hole and deposit a series of stainless steel ball along the viewline. AP and lateral stereotactic teleradiographs are obtained and these then serve as a reference mark for any possible shifts of the cranial contents during the course of the operation. If a shift is encountered, one simply moves the tumour in the three dimensional matrix to account for that so the updated images are correct. It has been our experience that such shifts are rare and when they do occur are small (Fig. 12).

Conclusion

The volumetric-stereotactic technique is superior to conventional freehand neurosurgical approaches to intra-axial lesions, where the surgeon's risk "getting lost" in attempts to find deep subcortical tumours and where the tissue planes between tumour and surrounding edematous brain tissue are unclear[4]. The stereotactic method maintains surgical orientation and thereby prevents loss of direction below the cortical surface[5]. Preoperatively data acquisition gathered by CT, MRI and angiographical examination gives an overview about the precise localization and size of the lesion and minimizes unnecessary brain tissue traumatization[1,2,5]. The procedure is of most benefit in deep-seated circumscribed lesions and posterior fossa lesions[3], including pilocytic juvenile astrocytomas, small vascular lesions, metastatic tumours and various localized miscellaneous lesions usually associated with complex partial seizures. The procedure is of less benefit in infiltrating tumours such as high grade gliomas and in fibrillary astrocytomas located in essential brain areas.

References

1. Goerss S, Kelly PJ, Kall B, *et al* (1982) A computed tomographic stereotactic adaption system. Neurosurgery 10: 375–379
2. Kelly PJ, Kall B, Goerss S *et al* (1983) Precision resection of intra-axial CNS lesions by CT-based stereotactic craniotomy and computer monitored CO_2-laser. Acta Neurochir (Wien) 68: ·1–9
3. Kelly PJ, Kall BA, Goerss S *et al* (1986) Computer-assisted stereotactic resection of posterior fossa lesions. Surg Neurol 25: 530–534
4. Kelly PJ (1986) Computer-assisted stereotaxis. Neurology 36: 535–541
5. Kelly PJ (1988) Volumetric stereotactic surgical resection of intra-axial brain mass lesions. Mayo Clin Proc 63: 1186–1198

Correspondence: A. Camacho, M.D., Ph.D., Department of Neurologic Surgery, Mayo Clinic, Rochester, MN 55905, U.S.A.

Acta Neurochirurgica, Suppl. 54, 89–92 (1992)
© by Springer-Verlag 1992

"Whole Body" Stereotaxy: Application of Stereotactic Endoscopy to Operations of Herniated Lumbar Discs

E.R. Heikkinen

Department of Neurosurgery, Oulu University Central Hospital, Oulu, Finland

Summary

In order to minimize surgical trauma of operations for lumbar disc prolapses, a modified Laitinen's stereotactic frame was utilized in nine patients. Stereotactic localization of the lesion could be done accurately in all cases. Percutaneous prolapsectomy and nucleotomy was performed successfully in three patients as a blind stereotactic procedure, whereas in the remaining six cases a microsurgical or conventional operation was needed to accomplish the removal of the prolapse or the discal mass.

Application of the endoscopic technique to lumbar disc operations is anticipated to make the percutaneous stereotactic prolapsectomy and nucleotomy a rational alternative to contemporary more invasive procedures.

Keywords: Stereotaxy; endoscopy; lumbar disc herniation; percutaneous nucletomy.

Introduction

Endoscopic operation technique has become an important part of various fields in general surgery. Also within neurosurgery several procedures are increasingly performed by using free hand endoscopy (Auer *et al.* 1988) or by combination of endoscopic and stereotactic techniques – especially operations within the ventricular system – (Bosch *et al.* 1978, Apuzzo *et al.* 1984, Heikkinen 1986). The idea of causing the least possible surgical trauma has been adopted for operations of lumbar disc prolapses by introducing microsurgical operation techniques (Caspar 1977), and even percutaneous procedures are possible for treating non-sequestrated lumbar disc protrusions or prolapses (Hijikata *et al.* 1975; Jacobson 1988).

The aim of the present work was to combine stereotactic localization techniques with percutaneous lumbar disc operations in order to make extirpation of sequestrated lumbar disc herniations possible.

Patients and Methods

Over 150 patients yearly are operated on for herniated lumbar disc at the Department of Neurosurgery of Oulu University Central Hospital. The indications for the operation include the sciatic syndrome with persisting clinical symptoms and signs in spite of conservative treatment for 1–3 months. The complaints of the patient must be consistent with a positive finding on computerized tomography (CT) examination and/or rhizography of the lumbar spine. During the year 1989, nine patients were included by informed consent in a trial of stereotactic prolapsectomy and nucleotomy. Preparatory measurements for the localization of the target, i.e. the sequestrated disc prolapse and the remaining intervertebral disc mass, were done using routine pre-operative CT pictures. As the gantry angle in the CT examination was written down, the plane of introduction of the various surgical instruments could be adjusted so as to be identical in the stereotactic operation. This was usually performed several days after the CT examination. For facilitating the orientation in the operation, an ink marking was drawn at the CT examination on the back of the patient in the mid-line, at the interspinous space at the level corresponding to the intervertebral level of the disc prolapse. An appropriate trajectory for the operation instruments was selected using this reference marker and intra-operative X-rays in antero-posterior and lateral projections. The trajectory had to go through the lamina, about 10 millimeters from the mid-line through the prolapsed mass, and to continue lateral to the nerve root into the corresponding intervertebral space.

A standard Laitinen's stereotactic frame (Issal Medical Products AB, Helsingborg, Sweden) was modified for making it mountable on a standard operation table as described earlier (Heikkinen ER, Proceedings of the X Congress of WSSFN, Stereotactic and Functional Neurosurgery, 1990, 54 + 55: 413 – 417). No bony fixation was needed although it is possible to arrange it by pushing the side cylinders of the stereoguide on to the iliac crests on both sides of the patient. Usually, the operation could be performed under local infiltration anaesthesia, the patient lying in the prone position (Fig. 1). A skin incision of less than one centimeter was made about 1.5 cm from the mid-line on the side of the prolapse. A percutaneous nucleotomy set (type Mayer and Brock, Aesculap, Tuttlingen, FRG) was utilized, consisting of a dilatator series, a trephine drill of 4 mm in diameter for the opening of the lamina, various forceps for extirpation of the prolapse mass, and a nucleotome for removing the remaining disc material. Several types of endoscopes including an Auer neuroendoscope were utilized mainly for surveying the end result of a microsurgical extirpation but not

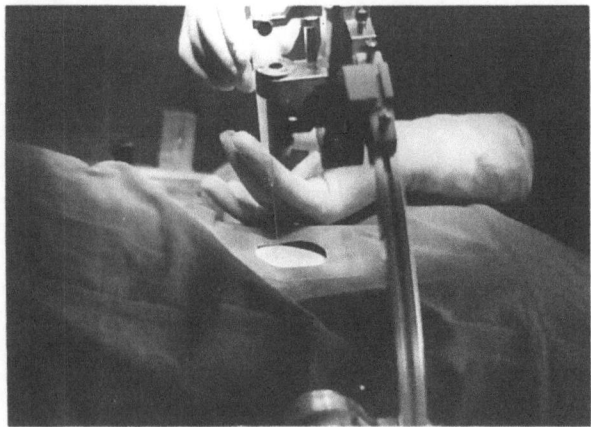

Fig. 1. Stereotactic frame in use for a lumbar operation. The patient is lying in the prone position draped with a sterile cover. Only local infiltration anaesthesia is needed around the entry point on the skin and along the trajectory of various instruments to the lamina

as a crucial part of the prolapsectomy. An example of a successful extirpation of a sequestrated prolapsed mass is shown in Fig. 2 (patient No 3 in Table 1).

Results

Stereotactic localization technique of both the herniated lumbar disc and the corresponding intervertebral space could be accurately done in all the nine cases. No root lesions, intra-operative haemorrhage, postoperative infections, or other complications were

encountered. In five out of the nine cases (Nos. 2, 3, 4, 5, and 8 in Table 1), the stereotactic extirpation of the sequestrated disc prolapse and the removal of the disc material from the intervertebral space could be done successfully under local anaesthesia. However, in two of these cases (Nos. 4 and 5, Table 1), the penetration of the lamina and manipulation close to the nerve root was painful, and therefore the operation was continued under general anaesthesia for ensure a complete removal of the disc material. The remaining three patients experienced an immediate relief of radicular pain intra-operatively after the removal of the prolapsed mass. In four cases (Nos. 1, 6, 7, and 9, Table 1) general anaesthesia was utilized, and in the other two patients (Nos. 2 and 9, Table 1) an open procedure was later on necessary because of recurrence of symptoms. All the patients could be mobilized on the first day after the operation. The stay in hospital was only a few days in those cases who underwent the percutaneous procedure, as the complaints caused by the surgery were minimal (Fig. 3). The duration of the postoperative sick-leave was 4–5 weeks as a routine in all the cases. The results are shown in Table 1.

Discussion

From our initial experiences using stereotactic localization techniques combined with percutaneous nucleotomy and endoscopic methods for operations of

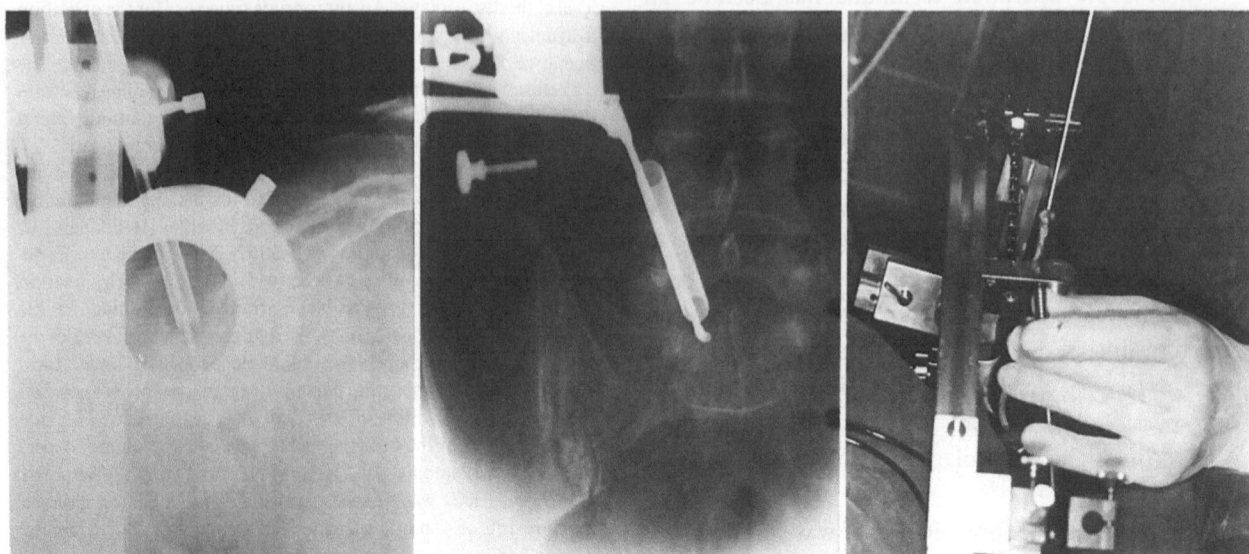

Fig. 2. Percutaneous stereotactic extirpation of a sequestrated disc prolapse. A dilatator set is guided exactly to a desired point on the lamina, which is then trephined by a drill, 4 mm in diameter. A sequestrated disc prolapse of 3.5 cm in length and 0.5 cm in breadth is removed in one piece. The remaining disc material can be extirpated by the nucleotome

Table 1. *Stereotactic Technique Applied to Operations for Herniated Lumbar Discs*

Pat.	Age	Localization	Anaesthesia	Operation	Result	Follow-up (mo)
1 M	27	pres. dx	G	stt-localization conventional ext.	exc.	12
2 M	59	LIV/V sin	L	stt-extirpation later conventional	exc.	9
3 M	37	pres. dx	L	stt-extirpation and nucleotomy	exc.	3
4 M	26	pres. dx	L + G	stt-nucleotomy and microsurg. ext.	exc.	2.5
5 M	43	pres. dx	L + G	stt-nucleotomy and microsurg. ext.	exc.	2
6 M	27	pres. dx	G	stt-localization, nucleotomy and conventional ext.	exc.	2
7 F	30	pres. dx	G	stt-nucleotomy	exc.	1.5
8 M	43	pres. dx	L	stt-nucleotomy	good	1
9 F	61	pres. sin	G	stt-nucleotomy, conventional ext.	exc.	1
		LIV/V dx	G	1 mo later		

pres. = presacral, M = male, F = female, G = general, L = local, ext. = extirpation, exc. = excellent.

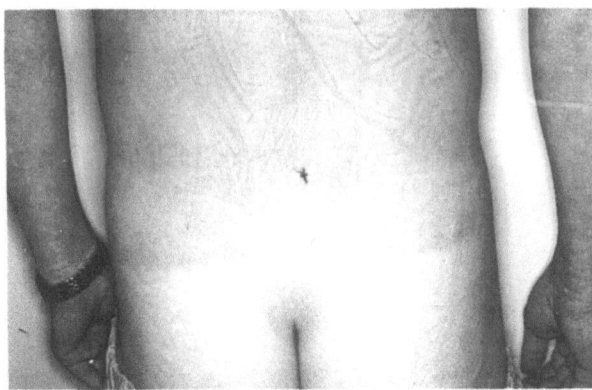

Fig. 3. Postoperative picture of a patient after a percutaneous stereotactic operation. The scar is less than 10 mm in length making early mobilization possible

herniated lumbar discs, the following drawbacks must be taken into account: 1. A sophisticated stereotactic frame (as well as routine) for performing stereotactic operations is necessary. 2. The stereotactic prolap-sectomy and nucleotomy using the percutaneous nucleotomy set is a blind procedure, where the completeness of the removal of the disc material cannot be exactly estimated. Therefore, recurrent herniation is possible. An intraoperative discography and use of compatible endoscopes with various working modalities including laser coagulation may help to overcome these difficulties in the near future.

The following advantages are evident even at the present stage of the development of the "whole body" stereotaxy: 1. Exact localization of targets in the lumbar area near the middle line is possible using a standard stereotactic frame; the instruments can be easily kept at a desired position on the instrument carrier of the stereo guide. 2. Because of the less invasive nature of the stereotactic percutaneous procedure, local infiltration anaesthesia may suffice even for removal of a sequestrated disc prolapse; complications seem to be minimal including post-surgical pain. 3. The stereotactic trial can be completed by a microsurgical or conventional operation when necessary.

In conclusion, stereotactic percutaneous operation technique is under development for use in lumbar disc surgery. The next step will be adoption of compatible endoscopes and intra-operative X-ray examinations for documenting the completeness of the extirpation. An operation on an out-patient basis may become possible and the sick-leave time is anticipated to be shortened due to the less invasive nature of the operative technique.

References

1. Apuzzo MLJ, Chandrasoma PT, Zelman V *et al* (1984) Computed tomographic guidance stereotaxis in the management of lesions of the third ventricular region. Neurosurgery 15: 502–508
2. Auer LM, Holzer P, Ascher PW, Heppner F (1988) Endoscopic neurosurgery. Acta Neurochir (Wien) 90: 1–14
3. Bosch DA, Rähn T, Backlund E-O (1978) Treatment of colloid

cysts of the third ventricle by stereotactic aspiration. Surg Neurol 9: 15–18

4. Caspar W (1977) A new surgical procedure for lumbar disc herniation causing less tissue damage through a microsurgical approach. Adv Neurosurg 4: 74–81

5. Heikkinen ER (1986) Stereotactic neurosurgery: New aspects of an old method. Annals Clin Res [Suppl] 47: 73–83

6. Hijikata S, Yamagishi M, Nakayama T et al (1975) Percuta-

neous discectomy: method for disc herniation. J Toden Hospital 5: 5–13

7. Jacobson S (1988) Lumbar percutaneous discectomy. Bull Hosp Dis Orthop Inst 48: 67–74

Correspondence: E. R. Heikkinen, M. D., Department of Neurosurgery, Oulu University Central Hospital, SF-90220 Oulu, Finland.

Acta Neurochirurgica, Suppl. 54, 93–97 (1992)

Potential Use of Robots in Endoscopic Neurosurgery

A.L. Benabid[1,3], **S. Lavallee**[2], **D. Hoffmann**[1,3], **P. Cinquin**[2], **J. Demongeot**[2], and **F. Danel**

[1]Unite INSERM U. 318, Neurobiologie Préclinique, Laboratoire de Neurobiophysique, UFR de Médecine, Université Joseph Fourier de Grenoble
[2]Laboratoire de Biomathematiques, UFR de Médecine, Université Joseph Fourier de Grenoble
[3]Service de Neurochirurgie at: CHU Albert Michallon, Grenoble, France

Summary

A 6-axis stereotactic robot has been designed and linked to a stereotactic frame for routine use. Robot software allows the positioning of a probe holder in order to reach a given target. A calibration step enables the robot to compute the position of the x-ray beam and correct its final position to avoid parallax errors. The co-ordinates of the target are presently taken from antero-posterior and lateral X-rays using a digitizing table. Connection with a digitized angiography system is in progress and will allow direct sampling of numerical data from the x-ray data. Further steps will include connections with a 3D-reconstructed image from MRI and CAT as well as with a resident computerized atlas. Present experience after 14 months of daily practice represents 140 stereotactic procedures which can be extended to any special use, including endoscopic approaches.

Keywords: Robot; stereotaxy; neurosurgery; computer assisted surgery.

Introduction

Due to the particular susceptibility of nervous tissue in the brain, there has always been a natural tendency to minimize the trauma of neurosurgical procedures including surgical approaches. This attitude culminates in the "keyhole neurosurgery" concept, which imposes two complementary prerequisites: precision of the target and narrowness of the approach. The second one has been made possible by technical advances in endoscopy, fiber optics, micro-instrumentation and the use of physical agents capable of being transported along thin pathways (such as heat and laser). The first prerequisite has been addressed for several years now by such methods as stereotaxy[28,29]. The increasing number of applications of stereotaxy, the availability of sophisticated and predigitized imaging modalities (CT-scan, MRI, DSA, US imaging) and the expanding power of computers call for integration of these technical opportunities[23]. Robotization of neurosurgical procedures is the logical and mandatory consequence of these developments. This was already recognized several years ago by Patrick Kelly[10-22] and, later, by Ronald Young[31] and Y.S. Kwoh[24-26]. All that remains to be accomplished[8], as well as the work presented here, is nothing more than the developments of their original and pioneering ideas.

The following is a current status report on a preliminary routine clinical application of a prototype robot designed to be integrated in a multi-purpose, universal system. It is, in the present state, mainly stereotactic but may easily be applicable to spinal surgery and to any other specific application which would appear useful.

Methods and Material

The robot presently in use is derived from a universal industrial robot[2,27] which has been redesigned for neurosurgical tasks. Basically, it is designed to be slower, since speed is not necessary, and as precise as possible. Maximal safety, which comes naturally from these two specifications, has also been given special attention.

The robot resembles a neurosurgeon and is similarly sized. It has a trunk, shoulder, arm, elbow, forearm, wrist and hand and all parts are connected by six angular joints. The fingers of the hand, which perform the skills, are momentarily restricted to a guide tube, which can be positioned to permit the introduction of a probe towards a target, as is usually the case during stereotaxy. Any other kind of "fingers" can easily be designed according to the requirements of the task. The most important feature confering human likeness to this robot is its "brain", provided by a micro-computer (IBM PC-AT), which mainly houses the software used to pilot the robot. Neurosurgeons, usually work in the brain as in a Cartesian space, where points are represented by the coordinates x, y, z, and a linear

Fig. 1. Robotized stereotactic set-up

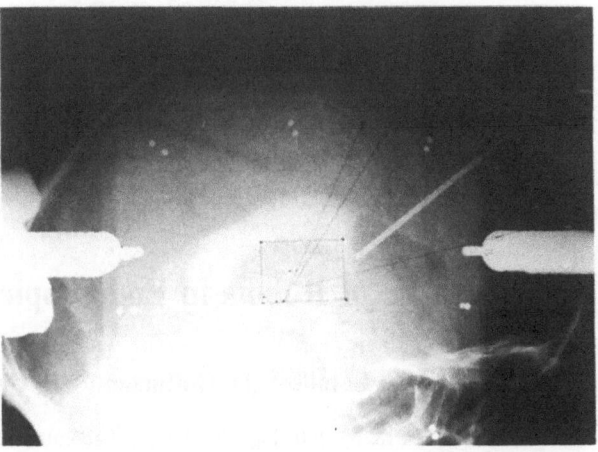

Fig. 2. X-ray appearance of the calibration cage and of the opaque beads, on a lateral X-ray where the target for thalamic implantation has been drawn. In this projection, the calibration cage beads, the head pins of the Talairach frame, and the target points will be digitized and their co-ordinates introduced into the computer which drives the robot

trajectory by two points. For the robot, the spatial coordinates of a point are a set of six different angles and even a linear trajectory between two points corresponds to a set of six complicated functions of time for its six angles. The changer matrix is a central feature of the robot's driving program. Feeding the robot's brain with the co-ordinates of these targets and trajectories can be achieved by different means. i) They can be keyed in as numerical values on the computer keyboard. This method can be used for small displacements of the robot, which are needed to correct, or even change, the final position of the guide tube. ii) They can also be digitized, using a digitizing table, from X-ray obtained during the stereotactic procedure. This method is usually used when the target and the primary trajectory are defined. iii) They could also be directly taken, using a "mouse" driven cursor, from the already digitized pictures provided by a digitized substraction angiography (DSA) system. This last feature is in the process of being added to our set-up and will by-pass the steps of taking X-rays, developing films and digitizing the points of interest. iv) Finally, they could be taken from 3D-reconstruction[6] of a spatial target, either a space occupying lesion (such as tumour, an abscess, a cyst, a haematoma, for biopsy or sampling) or a normal structure (such as a thalamic nucleus in the case of functional neurosurgery). Figure 1 shows the different parts of the set-up.

Results

1. General Operation

The first objective of the robot is to know where it is and where the patient held by the stereotactic frame is. To achieve this, a first subroutine has been written which tells the robot to place a calibrating cage around the head of the patient. This plexiglass cage looks like an open cubic box, the four sides surrounding the aperture each being implanted with nine X-ray opaque beads, the positions of which have been precisely measured. When this cage is placed around the head, two X-rays (antero-posterior and lateral projections) (Fig. 2) are taken. The beads appear on these films as two sets of nine dots, because of the parallax induced by the conicity of the X-rays beams, although the

radiogenic tubes are placed at a "long distance" (3.75 meters) from the center of the head. When the 18 dots have been digitized, the robot software calculates both the position of the central X-ray beam and the parallax parameters. This parallax error will then be automatically taken into account in all further calculations.

This first step is followed by stereotactic projective neuroradiological examinations, including angiography and ventriculography.

Angiography is performed when ever an invasive procedure is planned (biopsies, brachytherapy cannula implantation or SEEG recording depth electrodes) except if the trajectory is systematically placed in a cortical area known (and confirmed by practice) to be avascular, such as the precoronal area used for thalamic approaches. Ventriculography is also performed in order to provide the contours and anatomical landmarks of the third ventricle on which the atlas[30] references are based and which are used to calculate the thalamic targets of functional neurosurgery. Third ventricle anatomy will also be used to co-ordinate the stereotactic and extra-stereotactic images, such as those obtained from CT-scan and MRI pictures, which, in our set-up, are not yet possible in stereotactic conditions. To make them coherent, several methods can be used. For MRI, a set of adjacent parasagittal slices, including the midline plane, are taken. They are enlarged at the actual magnification coefficient of 1.05 for our stereotactic set-up. A grid is displayed on each magnified picture allowing them to be superimposed in order to obtain a final diagram containing all anatomical and pathological information provided by the various pictures. Among the anatomical data are features of the third ventricle (anterior, medial and posterior commissures, infundibulum, pre and post-mammillary incisurae) as well as the acqueduct of Sylvius and the fourth ventricle, which can then be applied and made coincident onto the lateral view of the stereotactic ventriculogram where these elements are also visible. It is therefore easy to report the position of pathological processes, mainly space occupying lesions on the stereotactic diagram.

For CT-scans[7,28], another method has been designed. The location of the slices is recognized on the basis of remarkable features (such as bony structures of the base of the skull, ventricular landmarks, calcified pineal gland, and choroid plexuses, injected vessels, etc). The average ratio between the distance from anterior to posterior inner tables of the skull as measured on the stereotactic lateral and frontal X-rays and on the corresponding CT-slices is calculated. The distance of the antero-posterior and lateral limits of space occupying lesions is measured on the CT-slices, multiplied by the magnification ratio, and reported on the stereotactic scheme to reconstruct the pattern of the lesion. The superimposition of all these different types of information constructs a final pattern on which the target of the trajectory is designed and calculated. The target, along with a second point, are digitized to define the trajectory. The resident computer software computes the six angle functions necessary to drive the robot from the stand-by to final positions, taking into account all forbidden places, such as the superstructures of the frame for instance.

Once the trajectory is computed, it is displayed and submitted to validation by the surgeon. When validation is given, the robot starts its approach. A few centimeters before reaching the final position, a new validation is required and the end of the approach is made at reduced speed. X-ray controls allow verification that the position reached is correct, and if not, correction can be made through a procedure of repeated small moves which can be triggered from the pilot computer keyboard. These moves can be performed in the Cartesian space of the frame or in the six dimensional angular space of the robot.

2. Clinical Applications

During the first 14 months of use in the stereotactic operating theatre, 140 procedures were performed, covering the whole range of routine applications encountered in current neurosurgical practice.

Tumour biopsies (94 cases) were usually performed using a lateral approach. In this situation, the software makes the probe holder co-linear with the local X-ray beam in order to avoid parallax errors. The control X-ray shows the projection of the trajectory as it will actually be in relation to the vessels, for instance, without any need for correction. For processes of the midline (colloid cysts or pineal gland tumours for instance), a double oblique approach was preferred. In these cases, it is easy to choose the best trajectory, for example passing through a foramen of Monro and ending at the center of the cyst or tumour. Positioning of a ventriculoscope would be significantly easier with the robot, particularly when several tracks are needed.

Functional neurosurgery, such as thalamic stimulation[4,5] (25 cases) for tremor or pain can also be easily performed by double oblique approaches based on anatomical landmarks calculated from third ventricle features. In this case, several tracks are often needed to provide an accurate definition of the target using electrophysiological methods, and the precise and quick placement of the robot is of considerable help. Other applications (21 cases), such as third ventriculostomy, stereo-electro-encephalographic electrode insertion (with up to 10 electrodes), and brachytherapy[3,9] with ^{125}I seeds or ^{192}Ir wires are also easily performed during current routine practice.

Discussion

The future of robotics in neurosurgery has to be carefully considered. As the forthcoming development of this methodology accelerates and as industrials become more and more interested in robot technology, the size of the market will stimulate progress in this field which in turn will open the way for new applications which cannot be performed at the present time due to technological limitations. All medical technologies have benefitted from this kind of process. Recent improvements in endoscopic methods have permitted new neurosurgical applications, but are presently limited in the accessibility of deep structures. The increased dexterity provided by robotization will make possible what is today considered utopian. Related fields such as engineering, optics, biomaterials, artificial vision, and miniaturization will contribute to the conception and realization of flexible robots, holding sensors (ultra-sonic, barometric or visual) capable of driving themselves along curved (and winding) trajectories amidst deep and fragile anatomical landscapes, such as the Sylvian fissures, the basal cisternae, and the ventricular cavities, towards deep seated, small sized or even moving targets.

At every step of this evolution, the double imperatives of decreasing invasiveness and increasing precision and efficiency must be kept in mind. Although robot technology will push back the frontiers of the feasable, the time has not yet come for imposing the three laws of robotics stated by Isaac Asimov[1]. These prerequisites, make a robot unable to harm a human being, oblige to obey human orders when they do not jeopardize the first law and commit it to protect itself as long as the first two laws are respected. Neither can a robot be heard asking why the surgeon bothered coming into the operating theatre. Despite our wildest neurosurgery-fiction fantasies, robots will be here for nothing else but helping and serving us.

As Dr. Mracek reminded me, the term robot comes from the Czechoslovakian word "robota" which means worker and is from the same root as "rabota" in Russian and "Arbeit" in German. As quoted by R. Young[31], it was used in the present sense for the first time in the science-fiction novel of K. Capek, Rossum's Universal Robots (Doubleday, New York, 1923). It reminds us that robots are not going to replace us, they are definitely designed only to work for us, depending on how well we program them.

References

1. Asimov I (1950) I, Robot. Doubleday, New York
2. Benabid AL, Cinquin P, Lavallée S, Le Bas JF, Demongeot J, de Rougemont J (1987a) Computer-driven robot for stereotactic surgery connected to CT scan and magnetic resonance imaging. Appl Neurophysiol 50: 153–154
3. Benabid AL, Chirossel JP, Mercier C, Louveau A, Passagia JG, Henry S, de Rougemont J, Vrousos C (1987b) Removable, adjustable and reusable implants for stereotactic interstitial radiosurgery of brain tumours. Appl Neurophysiol 50: 278–280
4. Benabid AL, Pollak P, Louveau A, Henry S, de Rougemont J, (1987c) Combined (thalamotomy and stimulation) stereotactic surgery of the VIM thalamus nucleus for bilateral Parkinson disease. Appl Neurophysiol 50: 344–346
5. Benabid AL, Pollak P, Hommel M, Gaio JM, de Rougemont J, Perret J (1989) Traitement du tremblement parkinsonien par stimulation chronique du noyau ventral intermediaire du Thalamus. Rev Neurol (Paris) 145(4): 320–323
6. Cinquin P (1987) Application des fonctions splines au traitement d'images numeriques. Thèse d'état de Sciences Mathématiques; Université Joseph Fourier, Grenoble
7. Colombo F, Angrilli F, Zanardo A, Pinna V, Alexandre A, Benedetti A (1982) A universal method to employ CT scanner spatial information in stereotactic surgery. Appl Neurophysiol 45: 352–354
8. Doll J, Schlegel W, Pastyr O, Sturm V, Maier-Borst W (1987) The use of an industrial robot as a stereotactic guidance system. CAR'79, 374–378
9. Gutin PH, Phillips TL, Wara WM, Leibel SA, Hosobuchi Y, Leven VA, Weaver KA, Lamb S (1984) Brachytherapy of recurrent malignant brain tumours with removable high activity iodine-125 sources. J Neurosurg 60: 61–68
10. Kall BA, Kelly PJ, Goerss SJ, Earnest F (1985b) IV. Cross-registration of points and lesion volumes from MR and CT. Proceed. 7° annual meeting of frontiers of engineering and computing in health care, pp 935–942
11. Kall BA, Kelly PJ, Goerss SJ (1985b) Interactive stereotactic surgery system for the removal of intracranial tumous utilizing the CO_2 laser and the CT-derived database. IEEE Trans Biomed Eng 32: 112–116
12. Kall BA, Kelly PJ, Goerss S (1987) Comprehensive computer-assisted data collection treatment planning and interactive surgery. Proceed. SPIE, Medical imaging 767: 509–514
13. Kall BA, Kelly PJ, Goerss SJ, Frieder G (1985c) Methodology and clinical experience with computed tomography and a computer-resident stereotactic atlas. Neurosurgery 17: 400–407
14. Kelly PJ, Kall BA, Goerss SJ (1984a) Transposition of volumetric information derived from computed tomography scanning into stereotactic space. Surg Neurol 21: 465–471
15. Kelly PJ, Alker GJ, Goerss S (1982a) Computer assisted stereotactic laser microsurgery for the treatment of intracranial neoplasms. Neurosurgery 10: 324–331
16. Kelly PJ, Alker GJ, Goerss S (1982b) Computer assisted laser microsurgery for the treatment of intracranial neoplasms. Neurosurgery 10: 324–331
17. Kelly PJ, Kall BA, Goerss S, Earnest F (1985) Present and future developments of stereotactic technology. Appl Neurophysiol 48: 1–6
18. Kelly PJ, Alker GJ (1980) A method for stereotactic laser microsurgery in the treatment of deep seated CNS neoplasms. Appl Neurophysiol 43: 210–215
19. Kelly PJ, Alker GJ (1981) A stereotactic approach to deep-seated central nervous system. Surg Neurol 15: 331–335
20. Kelly PJ, Alker GJ, Kall B, Goerss S (1984b) Method of computed-tomography based stereotactic biopsy with arteriographic control. Neurosurgery 14: 172–177
21. Kelly PJ (1986a) Technical approaches to identification and stereotactic reduction of tumour burden. In: Walker MD, Thomas DGT (eds) Biology of brain tumour. Martinus Nijhoff, Boston Dordrecht Lancaster, pp 237–343
22. Kelly PJ Kall B, Goerss S, Alker GJ (1983) Precision resection of intraaxial CNS lesions by CT-based stereotactic craniotomy and computer monitored CO_2 laser. Acta Neurochir (Wien) 68: 1–9
23. Kosugi Y, Watanabe E, Goto J, Watanabe T, Yoshimoto S, Takakura K, Ikebe J (1988) An articulated neurosurgical navigation system using MRI and CT images. IEEE Transactions on Biomedical Engineering 35(2): 147–152
24. Kwoh YS, Reed IS, Chen JY, Shao HM, Truong TK, Jonckheere EA (1985) New computerized tomographic-aided robotic stereotaxis system. Robotics Age 7: 17–22
25. Kwoh YS, Hou J, Jonckheere EA, Hayati S (1988) A robot with improved absolute positioning accuracy for CT guided stereotactic brain surgery. IEEE Transactions on Biomedical Engineering 35(2): 153–160
26. Kwoh YS, Young R (1990) Robotic aided surgery. In: Kelly PJ (ed) Computers in stereotactic neurosurgery. Blackwell Scientific Publication, Cambridge
27. Lavallée S (1989a) Gestes médico-chirurgicaux assistés par ordinateur: Thèse Sciences Mathématiques. Universite Joseph Fourier, Grenoble

28. Mundinger F, Birg W, Klar M (1978) Computer-assisted stereotactic brain operations by means including computerized axial tomography. Appl Neurophysiol 41: 169–182
29. Ostertag CB, Mennel HD, Kiessling M (1980) Stereotactic biopsy of brain tumours. Surg Neurol 14: 275–283
30. Schaltenbrand G, Wahren W (1977) Atlas for stereotaxy of the human brain, 2nd ed. Georg Thieme, Stuttgart
31. Young RJ (1987) Application of robotics to stereotactic neurosurgery. Neurological Research 9: 123–128

Correspondence: A.L. Benabid, M.D., Unite INSERM U.318, Neurobiologie Preclinique, Laboratoire de Neurobiophysique, UFR de Médecine, Université Joseph Fourier de Grenoble, CHU Albert Michallon, BP 277X, F-38043 Grenoble Cedex, France.

Subject Index

H.-J. Reulen, A. Baethmann, J. Fenstermacher,
A. Marmarou, M. Spatz (eds.)

Brain Edema VIII

(Acta Neurochirurgica / Supplementum 51)

1990. 203 figures. XIII, 416 pages.
Cloth DM 250,-, öS 1750,-
Reduced price for subscribers to "Acta Neurochirurgica":
Cloth DM 225,-, öS 1575,-
ISBN 3-211-82240-2

The book is a compilation of papers presented at the Eighth International Symposium on Brain Edema, held from June 17-20, 1990, in Bern, Switzerland. The Symposium explored clinical as well as basic science aspects of this topic. Clinicians and scientists from many fields of neurosciences contributed to the state-of-the art presentations and discussions.

The papers in this volume are grouped for the reader's convenience. Beside chapters on the pathophysiology the papers are focused around the major disease processes associated with brain edema: tumors, trauma, ischemia, hydrocephalus, hypertension, infection, pseudotumor cerebri etc. This allowed to group together the neuropathology, pathophysiology including cellular and molecular phenomena, clinical findings, in vivo diagnosis with MRI as well as therapeutic aspects of a given entity. A major issue was the role of biochemically active mediator substances in opening the blood brain barrier and development of edema. Methods of their specific inhibition may become the most important and effective therapeutic interventions in the future.

Editors, authors and Springer-Verlag have made a great effort to publish this volume – the eighth in the series – within 6 months after the Symposium. Thus, the volume reflects an up-to-date knowledge of the disease and the possible modes of treatment as well as new advances in innovative research. The book provides important information for all those who are clinically or scientifically involved with this disease.

E. Hitchcock, G. Broggi, J. Burzaco,
J. Martin-Rodriguez, B. Meyerson, S. Toth (eds.)

Advances in Stereotactic and Functional Neurosurgery 9

Proceedings of the 9th Meeting of the European Society for Stereotactic and Functional Neurosurgery, Malaga 1990

(Acta Neurochirurgica / Supplementum 52)

1991. 67 figures. Approx. 160 pages.
Cloth DM 160,-. öS 1120,-
Reduced price for subscribers to "Acta Neurochirurgica":
Cloth DM 144,-, öS 1008,-
ISBN 3-211-82283-6

The book gives the most up-to-date information for the expanding field of stereotactic and functional neurosurgery from European and International experts.

The newest developments in neural transplantation and stereotactic irradiation are included together with the reports on extensive trials of analgesic surgery and new techniques used in the treatment of a variety of functional disorders.

W. Koos, B. Richling (eds.)

Processes of the Cranial Midline

International Symposium, Vienna, Austria,
May 21-25, 1990

(Acta Neurochirurgica, Supplementum 53)

1991. 155 partly coloured figures. Approx. 200 pages.
Cloth DM 210,-, öS 1470,-
Reduced price for subscribers to "Acta Neurochirurgica":
Cloth DM 189,-, öS 1323,-
ISBN 3-211-82309-3

The editors have invited the leading neurosurgeons from around the world to discuss the treatment, approaches, anatomy, and results of pathological lesions that occur in the cranial midline. This region continues to challenge neurosurgeons and interventionalists to construct new and imaginative approaches and treatments in the continued quest to provide effective therapy with minimal morbidity and mortality.

Prices are subject to change without notice

Springer-Verlag Wien New York

Sachsenplatz 4–6, P.O. Box 89, A-1201 Wien · Heidelberger Platz 3, D-1000 Berlin 33
175 Fifth Avenue, New York, NY 10010, USA · 37-3, Hongo 3-chome, Bunkyo-ku, Tokyo 113, Japan